Insights into Game Theory

Few branches of mathematics have been more influential in the social sciences than game theory. In recent years, it has become an essential tool for all social scientists studying the strategic behavior of competing individuals, firms, and countries. However, the mathematical complexity of game theory is often very intimidating for students who have only a basic understanding of mathematics. *Insights into Game Theory* addresses this problem by providing students with an understanding of the key concepts and ideas of game theory without using formal mathematical notation. The authors use four very different topics (college admissions, social justice and majority voting, coalitions and cooperative games, and a bankruptcy problem from the Talmud) to investigate four areas of game theory. The result is a fascinating introduction to the world of game theory and its increasingly important role in the social sciences.

EIN-YA GURA is a Senior Lecturer at the Rothberg International School and a member of the Center for the Study of Rationality at the Hebrew University of Jerusalem.

MICHAEL B. MASCHLER is a Professor of Mathematics and a member of the Center for the Study of Rationality at the Hebrew University of Jerusalem. Professor Maschler is a world-renowned game theorist. He has had a long and fruitful collaboration with Robert Aumann (co-winner of the 2005 Nobel Prize for Economics), including their solution to a 2000-year-old puzzle on inheritance laws in the Talmud, discussed in this book.

Insights into Game Theory

An Alternative Mathematical Experience

EIN-YA GURA
AND
MICHAEL MASCHLER

CAMBRIDGE UNIVERSITY PRESS
Cambridge, New York, Melbourne, Madrid, Cape Town, Singapore,
São Paulo, Delhi

Cambridge University Press
The Edinburgh Building, Cambridge CB2 8RU, UK

Published in the United States of America by Cambridge University Press,
New York

www.cambridge.org
Information on this title: www.cambridge.org/9780521696920

First published 2008

Printed in the United Kingdom at the University Press, Cambridge

A catalogue record for this publication is available from the British Library

Library of Congress Cataloguing in Publication data

Gura, Ein-ya.
Insights into game theory : an alternative mathematical experience /
Ein-Ya Gura and Michael Maschler.
p. cm.
Includes bibliographical references and index.
ISBN 978-0-521-87422-9 (hardback)
1. Game theory. I. Maschler, Michael, 1927– II. Title.
QA269.G83 2008
519.3–dc22 2008023583

ISBN 978-0-521-87422-9 hardback
ISBN 978-0-521-69692-0 paperback

This book is dedicated to the memory of Michael Maschler,
who passed away on July 20, 2008.

Contents

Preface

This book is a *tour de force* of what may be called "verbal mathematics." It demonstrates conclusively that mathematics is not a matter of symbols and equations; rather, it may be characterized as "precise reasoning that has considerable depth, complexity, or sophistication." The book is accessible to everyone who can think.

Also, it is a wonderful introduction to game theory; rather than "explaining" what the theory is about, it simply *does* it. If somebody came from Mars and wanted to know what we mean by "music," you could try to "explain" it; but it would be better to play a Bach fugue, a Verdi aria, some Louis Armstrong jazz, and "Lucy in the Sky with Diamonds." The second alternative is what Gura and Maschler do. Enjoy!

Robert J. Aumann

Introduction

Game theory is a relatively young branch of mathematics that goes back to the publication of *Theory of Games and Economic Behavior* by John von Neumann and Oskar Morgenstern in 1944.[1]

Game theory undertakes to build mathematical models and draw conclusions from these models in connection with interactive decision-making: situations in which a group of people not necessarily sharing the same interests are required to make a decision.

The choice of the topics reflects our purpose: we wanted to present material that does not require mathematical prerequisites and yet involves deep game-theoretic ideas and some mathematical sophistication. Thus, we ruled out topics from non-cooperative game theory, which requires some knowledge of probability, matrices, and point-set topology.

Broadly speaking, the topics chosen are all related to the various meanings that can be given to the concept of "fair division." The four chapters illustrate this.

The first, "Mathematical Matching," concerns, among other things, the problem of assigning applicants to institutions of higher learning. Each applicant ranks the universities according to his scale of preferences. The institutions of higher learning, in turn, rank the applicants for admission according to their own scale of preferences. The question is how to effect the "matching" between the applicants and the universities. The reader will discover that this problem leads to unexpected solutions.

The second chapter, "Social Justice," concerns social decision rules. In a democratic society it is customary to make decisions by

[1] Several "game-theoretic" topics had been discussed before this publication, but not in any systematic way.

a vote. The decision supported by the majority of voters is adopted. But the reader will discover that "majority rule" does not always yield clear-cut solutions. The attempt to find other voting rules raises unexpected difficulties.

The third chapter, "The Shapley Value in Cooperative Games," addresses, among other things, the following problem: a group of people come before an arbitrator and inform him of the expected profits of every subgroup, as well as of the whole group, if the groups operate independently. It seems that these data are sufficient for the arbitrator to decide how to divide the profits if all the litigants operate jointly.

The fourth chapter, "Analysis of a Bankruptcy Problem from the Talmud," addresses the following problem: several creditors have claims to an estate, but the total amount of the claims exceeds the value of the estate. How should the estate be divided among the creditors? In the chapter several solutions are accepted, two of which are discussed in the Talmud.

As explained above, this book is not a textbook in game theory. Rather, it is a collection of a few topics from the theory intended to open a window onto a new and fascinating world of mathematical applications to the social sciences. Our hope is that it will motivate the reader to take a solid course in game theory.

One of the aims of the book is to acquaint the reader and the student with "a different mathematics" – a mathematics that is not buried under complicated formulas, yet contains deep mathematical thinking. Another aim is to show that mathematics can efficiently handle social issues. A third aim is to deepen the mathematical thinking of the person who studies this book.

We believe that by studying the topics of this book, the mathematical thinking of the student will be enriched.

This book selects a small number of topics and studies them in depth. It shows the student of the social sciences how a mathematical model can be constructed for real-life issues.

The chapters are independent. A teacher and a student can choose one chapter or several and cover them in any order.

In high schools, the book can be used by students on any program track or as extracurricular material. The teacher can proceed to the deeper parts of each chapter if she has a mathematically inclined class or skip some of the proofs if the class cannot handle them. The book can also be used by students who want to read independently or under the guidance of a teacher beyond what is required in school.

At universities and colleges, the book can be used in courses whose aim is to introduce general game-theoretic topics and deepen mathematical thinking.

This book owes its origin to the PhD thesis of co-author Ein-Ya Gura. We thank the Science Teaching Center at the Hebrew University of Jerusalem for permission to publish this translation from Hebrew, the Center for the Study of Rationality for funding the translation, and the translator, Michael Borns, not only for the accuracy of his translation, but for the competence of his editing. We thank James Morrow for taking the time to read and comment on the manuscript, Zur Shapira for recommending it to Cambridge University Press, and Chris Harrison and the staff of Cambridge University Press for their encouragement and help in bringing the book to its final form. Last, but not least, we thank Robert Aumann for providing the impetus for both the Hebrew and the English publication of this book.

Ein-Ya Gura and Michael Maschler,
The Hebrew University of Jerusalem,
April 2008

1 Mathematical Matching

1.1 INTRODUCTION

In 1962 a paper by David Gale and Lloyd S. Shapley[1] appeared at the RAND Corporation, whose title, "College Admissions and the Stability of Marriage," raised eyebrows. Actually, the paper dealt with a matter of some urgency.

According to Gale,[2] the paper owes its origin to an article in the *New Yorker*, dated September 10, 1960, in which the writer describes the difficulties of undergraduate admissions at Yale University. Then as now, students would apply to several universities and admissions officers had no way of telling which applicants were serious about enrolling. The students, who had every reason to manipulate, would create the impression that each university was their top choice, while the universities would enroll too many students, assuming that many of them would not attend. The whole process became a guessing game. Above all, there was a feeling that actual enrollments were far from optimal.

Having read the article, Gale and Shapley collaborated. First, they defined the concept of stable matching, and then proved that stable matching between students and universities always exists. This and further developments will be discussed in this chapter.

For simplicity, Gale and Shapley started with the unrealistic case in which there are exactly n universities and n applicants and each university has exactly one vacancy. A more realistic description of this case is a matching between men and women – hence the title of their paper.

[1] Gale, D. and Shapley, L. S. 1962. "College admissions and the stability of marriage," American Mathematical Monthly 69: 9–15.

[2] Gale, D. 2001. "The two-sided matching problem: origin, development and current issues," International Game Theory Review 3: 237–52.

1.2 THE MATCHING PROBLEM

Consider a community of men and women where the number of men equals the number of women.

Objective: Propose a good matching system for the community.[3] To be able to propose such a system, we shall need relevant data about the community. Accordingly, we shall ask every community member to rank members of the opposite sex in accordance with his or her preferences for a marriage partner. We shall assume that no man or woman in the community is indifferent to a choice between two or more members of the opposite sex.[4] For example, if Al's list of preferences consists of Ann, Beth, Cher, and Dot, in that order, then Al ranks Ann first, Beth second, Cher third, and Dot fourth.[5] Again, we shall assume that Al is not indifferent to a choice between two or more of the four women on his list.

Example:
The men are Al, Bob, Cal, Dan.
The women are Ann, Beth, Cher, Dot.
Their list of preferences is:

Women's Preferences:

	Ann	Beth	Cher	Dot
Al	1	1	3	2
Bob	2	2	1	3
Cal	3	3	2	1
Dan	4	4	4	4

Men's Preferences:

	Ann	Beth	Cher	Dot
Al	3	4	1	2
Bob	2	3	4	1
Cal	1	2	3	4
Dan	3	4	2	1

Explanation: The numbers in the table indicate what rank a man or woman occupies in the order of preferences. For example, according to the men's ranking of the women, Al ranks Cher first, Dot second,

[3] The meaning of "good" will become clear presently.

[4] This assumption is introduced to simplify our task. In Section 1.10 we shall see how to dispense with it.

[5] If Al prefers Ann to Beth and Beth to Cher, it follows that he prefers Ann to Cher. Accordingly, we may list all his preferences in a row.

Ann third, and Beth last. And according to the women's ranking of the men, Cher ranks Bob first, Cal second, Al third, and Dan last. Thus Al ranks Cher first, while Cher ranks Al just third. If we pair them off, the match will not work out, if the first or second candidate on Cher's preference list agrees to be paired off with her.

Given everyone's preferences, can you propose a matching system for the community?

A Possible Proposal:

$$\begin{pmatrix} \text{Al} & \text{Bob} & \text{Cal} & \text{Dan} \\ | & | & | & | \\ \text{Dot} & \text{Ann} & \text{Beth} & \text{Cher} \end{pmatrix}$$

$$2 \times 2 \quad 2 \times 2 \quad 2 \times 3 \quad 2 \times 4$$

The numbers below each couple indicate what rank one member of a couple assigns to the other member. The number on the left indicates what rank the man assigns to the woman; the number on the right, what rank the woman assigns to the man. (Verify it!)

Argument for the Proposal:

(1) No members of any couple rank each other first.

(2) No members of any couple rank each other 1×2 or 2×1.

(3) The members of two couples rank each other second.

(4) Cal can be paired off with Cher or Beth, but he prefers Beth.

(5) That leaves Dan and Cher, who can be paired off.

This is indeed a possible proposal, but it is not a good one.

Cher is displeased, because she is paired off with her last choice. She can propose to Bob, but she will be turned down because she is his last choice. She will fare no better with Cal, because she is his third choice while he is paired off with his second choice. On the other hand, if Cher proposes to Al, he will be very pleased, because she is his first choice.

The proposal is rejected, because Cher and Al prefer each other to their actual mates, and one can reasonably assume that they will reject the matchmaker's proposal.

Another Possible Proposal: Let us try to pair off all the men with their first choice.

Al's first choice is Cher.
Bob's first choice is Dot.
Cal's first choice is Ann.
Dan's first choice is Dot.

We see that there is a problem: both Bob and Dan prefer Dot. We can try to pair off Dan with his second choice, Cher, but she is already paired off with Al. Will Dan's third choice work out? Dan's third choice is Ann, but she is already paired off with Cal. That leaves Dan with his last choice, Beth.

$$\begin{pmatrix} \text{Al} & \text{Bob} & \text{Cal} & \text{Dan} \\ | & | & | & | \\ \text{Cher} & \text{Dot} & \text{Ann} & \text{Beth} \end{pmatrix}$$
$$1 \times 3 \quad 1 \times 3 \quad 1 \times 3 \quad 4 \times 4$$

Three of the four men are paired off with their first choice. Do you think this proposal will be accepted or rejected?

Still Another Possible Proposal: Now we shall try to pair off all the women with their first choice. Is it possible?

Ann's first choice is Al.
Beth's first choice is Al.
Cher's first choice is Bob.
Dot's first choice is Cal.

We see that if we pair off Ann with her first choice, Al, then Beth cannot be paired off with him too. We can pair off Beth with her second choice, Bob, but he is already paired off with Cher. And Beth's third choice, Cal, is already paired off with Dot. Beth is therefore left with her last choice, Dan.

The new matching system is:

$$\begin{pmatrix} \text{Ann} & \text{Beth} & \text{Cher} & \text{Dot} \\ | & | & | & | \\ \text{Al} & \text{Dan} & \text{Bob} & \text{Cal} \end{pmatrix}$$

$$3 \times 1 \quad 4 \times 4 \quad 4 \times 1 \quad 4 \times 1$$

Three of the four women are paired off with their first choice. Will they accept or reject this matching system?

Beth can fight this matching. For example, she can approach Bob and suggest that they both reject this matching and form their own pair. In so doing Beth gets her second choice – better than her fourth choice – and Bob gets his third choice – better than his fourth choice. Thus, the above matching will be rejected by Beth and Bob.

Exercise: Analyze the second proposal above and see whether it can be rejected by any pair of men and women.

The first proposal was rejected, but we can turn the failed effort to our advantage. Indeed, we have learned that a matching system must satisfy the following requirement:

A matching system must be such that under it there cannot be found a man and a woman who are not paired off with each other but prefer each other to their actual mates.

Explanation: The matching system must be such that under it Ms. X cannot be paired off with Mr. x and Ms. Y cannot be paired off with Mr. y, when Ms. X prefers Mr. y to Mr. x and Mr. y prefers Ms. X to Ms. Y.

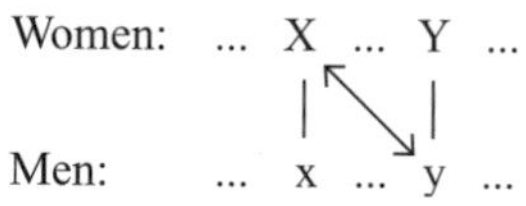

Women: ... X ... Y ...
| ↖↘ |
Men: ... x ... y ...

The figure indicates the "impossible" part of the system. Specifically, the double arrow indicates that X prefers y to x and y prefers X to Y.

If couples X–x and Y–y were paired off according to the matchmaker's recommendation, then Ms. X could say to Mr. y, "You prefer me to your actual mate and I prefer you to mine. Let's leave them and pair up."

Discussion:

Will Ms. X and Mr. y pair themselves off with each other? Not necessarily! Mr. y might say, "Yes, I prefer you, X, to Y, but I prefer Z to you."

If y is lucky and Z prefers him to her actual mate, then those two can pair themselves off with each other. Otherwise, y's best choice will be X, whom he prefers to his actual mate. *Either way, the matchmaker's recommendation will not be implemented.*

Definition: A matching system is called *stable* if under it there cannot be found a man and woman who are not paired off with each other but prefer each other to their actual mates.

Example:

For simplicity, we substitute letters for names.

The men: a, b, c, d.

The women: A, B, C, D.

Preference Structure:

	A	B	C	D
a	1	2	4	②
b	②	4	2	1
c	3	①	1	3
d	4	3	③	4

	A	B	C	D
a	4	2	1	③
b	②	1	3	4
c	3	①	4	2
d	2	4	①	3

We have circled a stable matching system in the above preference structure. Later we shall learn how to find such a system.

$$\begin{pmatrix} A & B & C & D \\ | & | & | & | \\ b & c & d & a \end{pmatrix}$$

$2 \times 2 \quad 1 \times 1 \quad 1 \times 3 \quad 3 \times 2$

Note: The position of the circles in the two tables must be identical.

Verification: Mr. c and Mr. d are paired off with their first choice, so they need look no further. Mr. b prefers Ms. B to his actual mate, Ms. A, but Ms. B will turn him down because he is her last choice. Mr. a prefers Ms. B and Ms. C to his actual mate, D. If he proposes to B, she will turn him down because he is her second choice and she is paired off with her first choice. If he proposes to C, she too will turn him down because he is her last choice.

Remark: When no man wants to deviate from the matchmaker's recommendation, then it does not matter if a woman wants to change, because she will not find a man who will agree to cooperate with her. Thus, there is no further need to continue the verification.

1.3 EXERCISES

1. Given the following preference structure, check whether the proposed matching systems are stable. Support your answer.

Women: A, B, C. Men: a, b, c.

Women's Preferences:

	A	B	C
a	1	1	1
b	2	2	2
c	3	3	3

Men's Preferences:

	A	B	C
a	1	2	3
b	1	2	3
c	1	2	3

i.

$$\begin{pmatrix} A & B & C \\ | & | & | \\ a & b & c \end{pmatrix}$$

ii.

$$\begin{pmatrix} A & B & C \\ | & | & | \\ a & c & b \end{pmatrix}$$

2. Given the following preference structure, check whether the proposed matching systems are stable.

Women: A, B, C. Men: a, b, c.

Women's Preferences:

	A	B	C
a	2	2	1
b	1	3	3
c	3	1	2

Men's Preferences:

	A	B	C
a	1	2	3
b	1	2	3
c	1	2	3

i.

$$\begin{pmatrix} A & B & C \\ | & | & | \\ a & b & c \end{pmatrix}$$

ii.

$$\begin{pmatrix} A & B & C \\ | & | & | \\ c & a & b \end{pmatrix}$$

iii.

$$\begin{pmatrix} A & B & C \\ | & | & | \\ b & a & c \end{pmatrix}$$

iv.

$$\begin{pmatrix} A & B & C \\ | & | & | \\ b & c & a \end{pmatrix}$$

3. Given the following preference structure of a community of four men and four women:

Women: A, B, C, D. Men: a, b, c, d.

Women's Preferences:

	A	B	C	D
a	1	2	4	2
b	2	4	2	1
c	4	1	1	3
d	3	3	3	4

Men's Preferences:

	A	B	C	D
a	4	2	1	3
b	2	1	3	4
c	3	1	4	2
d	2	4	1	3

(1) Is the matching system $\begin{pmatrix} A & B & C & D \\ | & | & | & | \\ b & c & d & a \end{pmatrix}$ stable?

If so, explain. If not, indicate which couple(s) will not follow the recommendation.

(2) Is the matching system $\begin{pmatrix} A & B & C & D \\ | & | & | & | \\ b & a & d & c \end{pmatrix}$ stable?

If so, explain. If not, indicate which couple(s) will not follow the recommendation.

(3) Using the above preference structure, propose a possible matching system for this community and check its stability.

4. Given the following preference structure of a community of five women and five men:

Women: A, B, C, D, E. Men: a, b, c, d, e.

Women's Preferences:

	A	B	C	D	E
a	5	4	3	2	1
b	1	5	4	3	2
c	2	1	5	4	3
d	3	2	1	5	4
e	4	3	2	1	5

Men's Preferences:

	A	B	C	D	E
a	1	2	3	4	5
b	5	1	2	3	4
c	4	5	1	2	3
d	3	4	5	1	2
e	2	3	4	5	1

(1) Show that the following matching systems are all stable.

i.
$$\begin{pmatrix} A & B & C & D & E \\ | & | & | & | & | \\ a & b & c & d & e \end{pmatrix}$$

ii.
$$\begin{pmatrix} A & B & C & D & E \\ | & | & | & | & | \\ e & a & b & c & d \end{pmatrix}$$

iii.
$$\begin{pmatrix} A & B & C & D & E \\ | & | & | & | & | \\ d & e & a & b & c \end{pmatrix}$$

iv.
$$\begin{pmatrix} A & B & C & D & E \\ | & | & | & | & | \\ c & d & e & a & b \end{pmatrix}$$

(2) Find another matching system with a similar structure. Is it stable?

(3) Verify that in this preference structure all preferences of the women are the reverse of the preferences of the men. For example, Ms. A is Mr. a's first preference, while Mr. a is Ms. A's last preference.

1.4 FURTHER EXAMPLES

In this section we shall present several preference structures and check whether there are any stable matching systems.

Example 1

The preference structure is:

	A	B
a	1	1
b	2	2

	A	B
a	1	2
b	1	2

There are two possible matching systems for a community of two men and two women. Let us check whether they are stable.

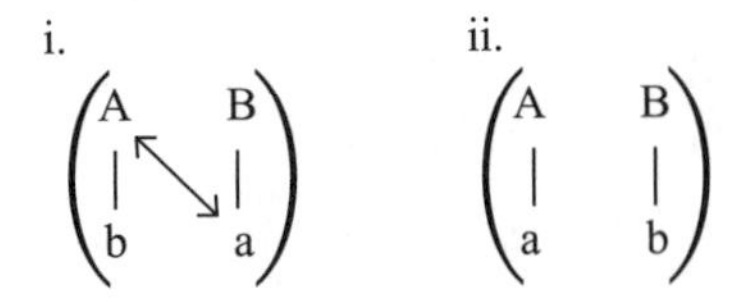

i. This matching system is unstable. The double arrow shows how the system can be undermined.
ii. This matching system is stable because A and a are paired off with their first choice and therefore will not deviate from the matchmaker's recommendation.

Example 2

The preference structure is:

	A	B
a	2	1
b	1	2

	A	B
a	1	2
b	1	2

The possible matching systems are:

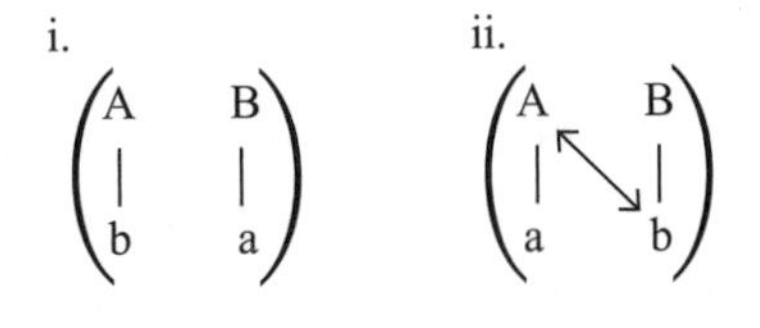

i. This matching system is stable because A and b are paired off with their first choice and therefore will not deviate from the matchmaker's recommendation.
ii. This matching system is unstable because A and b are paired off with their second choice and if they pair themselves off with each other, they will be better off.

Example 3

The preference structure is:

	A	B
a	2	1
b	1	2

	A	B
a	1	2
b	2	1

The possible matching systems are:

i.
$$\begin{pmatrix} A & B \\ | & | \\ b & a \end{pmatrix}$$

ii.
$$\begin{pmatrix} A & B \\ | & | \\ a & b \end{pmatrix}$$

i. This matching system is stable because A and B are paired off with their first choice and therefore will refuse to annul the match.
ii. This matching system is stable, too, because a and b are paired off with their first choice and therefore will refuse to deviate from the matchmaker's recommendation.

This example shows that if one side (say, the men's) obtains its first choice, that is, if one side is satisfied with the match, then the other side cannnot undermine the system, however dissatisfied it may be with the match.

A stable matching system is not necessarily a system under which everyone is satisfied. A matching system is stable when no unmatched pair will find it beneficial to deviate from the matching and form their own match. In other words, a stable matching system serves the interests of the matchmaker, whose recommendation will be honored, but it does not necessarily serve the interests of all community members.

Example 4

The preference structure is:

	A	B	C
a	3	2	1
b	1	3	2
c	2	1	3

	A	B	C
a	1	2	3
b	3	1	2
c	2	3	1

Two stable matching systems are straightforward in this example:

i. The men obtain their first choice: $\begin{pmatrix} A & B & C \\ | & | & | \\ a & b & c \end{pmatrix}$.

ii. The women obtain their first choice: $\begin{pmatrix} A & B & C \\ | & | & | \\ b & c & a \end{pmatrix}$.

There is also a third stable matching system, where men and women all obtain their second choice: $\begin{pmatrix} A & B & C \\ | & | & | \\ c & a & b \end{pmatrix}$.

In fact, all those who try to obtain their first choice will be rejected because they themselves are their favorite's third choice. (Verify it!)

There are three other possible matching systems for the given preference structure (see below). Show that they are all unstable.

i. $\begin{pmatrix} A & B & C \\ | & | & | \\ c & b & a \end{pmatrix}$ ii. $\begin{pmatrix} A & B & C \\ | & | & | \\ b & a & c \end{pmatrix}$ iii. $\begin{pmatrix} A & B & C \\ | & | & | \\ a & c & b \end{pmatrix}$

Explain why there is no other matching system.

Example 5: The Roommate Problem

We need to divide an even-numbered set of boys into pairs of roommates. Here again, a set of pairs will be called stable if in it there cannot be found two boys who are not roommates but prefer each other to their actual roommates. We shall see that a stable division does not always exist. In the following example there are three ways of dividing the boys into pairs. But in each case the division is unstable.

	a	b	c	d
a	–	1	2	3
b	2	–	1	3
c	1	2	–	3
d	1	2	3	–

The arrows indicate the pairs likely to undermine the division.

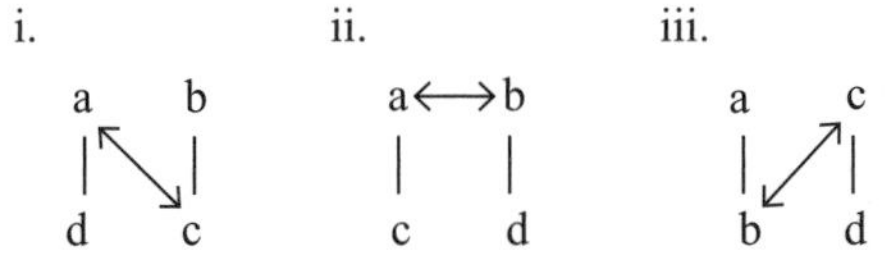

i. a b d c

ii. a⟷b c d

iii. a c b d

In this example d is everyone's last choice, and therefore his roommate will look for another roommate. In each case d's roommate will find a partner who likes him as his first choice, and therefore this partner will agree to deviate from the proposed division.

To sum up, the last example shows that in the case of the roommate problem it is possible that no stable matching exists. On the other hand, all the examples of the marriage problem provided thus far have had a stable matching. Whether this is always the case in marriage problems is not evident and requires either a proof or a counter-example.

One can see the difference between the two cases: in the roommate problem you can pair any two members whereas in the marriage problem you cannot pair any two members: namely, you cannot pair any two women and you cannot pair any two men.

1.5 EXERCISES

1. Is there a stable division of the set into pairs, given the following preference structure of the boys? If so, provide an example. If not,

show that no stable matching system exists by checking all the possibilities.

	a	b	c	d
a	–	1	2	3
b	3	–	1	2
c	1	2	–	3
d	2	3	1	–

2. Is there a stable division into pairs, given the following preference structure? If so, provide an example. If not, show that no stable matching system exists by checking all the possibilities.

	a	b	c	d
a	–	3	1	2
b	2	–	1	3
c	2	3	–	1
d	1	3	2	–

3. Is there a stable division into pairs, given the preference structure below? If so, provide an example.

	a	b	c	d
a	–	2	3	1
b	1	–	3	2
c	3	2	–	1
d	1	2	3	–

4. Is there a stable division into pairs, given the following preference structure? If so, provide an example.

	a	b	c	d
a	–	1	3	2
b	2	–	1	3
c	2	1	–	3
d	2	3	1	–

1.6 A PROCEDURE FOR FINDING STABLE MATCHING SYSTEMS (THE GALE–SHAPLEY ALGORITHM)

In connection with a procedure for finding stable matching systems, three questions naturally arise:

1. Given a preference structure, is one always certain of finding a stable matching system? Let us recall that a stable division does not always exist for the roommate problem. Does a stable division always exist for the matching problem?
2. How does one find a stable matching system?
3. Does only one stable matching system exist? (The answer is no; we have already seen examples where more than one stable matching system exists.)

In this section we shall answer the first question in the affirmative, using a procedure that terminates in a stable matching system. In so doing, we shall be in a position to answer also the second question.

The Gale–Shapley algorithm for finding a stable matching system

First Stage: Every man turns to the woman who is first on his list and proposes to her. Every woman who receives more than one proposal selects her favorite from among those who propose to her and tells the others that she will never marry them. Every man who is not rejected is put on a "waiting list" of the woman to whom he proposed.

Second Stage: Every man who was rejected turns to the woman who is second on his list and proposes to her. Every woman who receives more than one proposal, including any proposals from the previous stage, selects her favorite, and puts him on her waiting list. She informs the others that they are rejected.

Third Stage: Every man who is rejected turns to the woman who is next on his list – the second on his list if he was put on the waiting list at a previous stage, or the third on his list if he was rejected twice. Once again, every woman selects her favorite from among those who have proposed to her, including anyone on her waiting list from the previous stage, puts him on the waiting list, and rejects the others.

The procedure continues – until such time as no man is rejected. At that stage every man on a waiting list becomes a mate, and the procedure terminates.

Later we shall prove that this procedure leads to a stable matching system.

Example 1

The preference structure is:

	A	B	C	D
a	3	3	2	3
b	4	1	3	2
c	2	4	4	1
d	1	2	1	4

	A	B	C	D
a	1	2	3	4
b	1	4	3	2
c	2	1	3	4
d	4	2	3	1

Stage 1:

A	B	C	D
a	c		d
b*			

Mr. b, marked with an asterisk (*), is rejected. The others are "waitlisted."

Stage 2:

A	B	C	D
a	c		d*
			b

Mr. d is rejected (despite having been waitlisted at the previous stage).

Stage 3:

A	B	C	D
a	c*		b
	d		

Mr. c is rejected.

Stage 4:

A	B	C	D
a*	d		b
c			

Mr. a is rejected.

Stage 5:

A	B	C	D
c	d		b
	a*		

Mr. a is rejected again.

Stage 6:

A	B	C	D
c	d	a	b

The procedure ends and the proposed matching system is:

$$\begin{pmatrix} A & B & C & D \\ | & | & | & | \\ c & d & a & b \end{pmatrix}.$$

Exercise: Show that the above matching system is stable.

Example 2

The preference structure is:

	A	B	C	D
a	1	3	4	3
b	2	1	2	4
c	3	2	3	1
d	4	4	1	2

	A	B	C	D
a	1	2	3	4
b	1	2	4	3
c	1	2	4	3
d	4	1	2	3

Stage 1:

A	B	C	D
a	d		
b*			
c*			

b and c are rejected.

Stage 2:

A	B	C	D
a	d*		
	b		
	c*		

c and d are rejected.

Stage 3:

A	B	C	D
a	b	d	c

The procedure ends and the proposed matching system is:

$$\begin{pmatrix} A & B & C & D \\ | & | & | & | \\ a & b & d & c \end{pmatrix}.$$

1.7 EXERCISES

1. Find a stable matching system for the following preference structure:, using the Gale–Shapley algorithm, when the men propose to the women:

Women: A, B, C, D.

	A	B	C	D
a	1	2	4	2
b	2	4	2	1
c	3	1	1	3
d	4	3	3	4

Men: a, b, c, d.

	A	B	C	D
a	4	2	1	3
b	2	1	3	4
c	3	1	4	2
d	2	4	1	3

2. Find a stable matching system for the following preference structure, using the Gale–Shapley algorithm, when the women propose to the men:

Women: A, B, C.

	A	B	C
a	3	2	3
b	2	3	1
c	1	1	2

Men: a, b, c.

	A	B	C
a	1	2	3
b	1	2	3
c	2	1	3

3. Find a stable matching system for the following preference structure (1) when the women propose to the men, and (2) when the men propose to the women:

Women: A, B, C, D, E.

	A	B	C	D	E
a	5	4	4	5	1
b	4	5	5	4	2
c	3	2	3	2	3
d	2	3	1	1	4
e	1	1	2	3	5

Men: a, b, c, d, e.

	A	B	C	D	E
a	1	2	3	4	5
b	1	2	4	3	5
c	2	1	3	5	4
d	4	2	1	5	3
e	4	3	2	1	5

1.8 A STABLE MATCHING SYSTEM ALWAYS EXISTS

In this section we shall prove that for any preference structure, there is at least one stable matching system. We shall prove this in two stages. First, we shall show that the procedure described in Section 1.6 must terminate after a finite number of steps. Second, we shall show that the final step in the procedure must be a stable matching system.

Theorem:

The Gale–Shapley algorithm terminates after a finite number of steps.

Discussion:

Before proceeding to the proof, let us mention why the theorem is necessary. It might seem, at first glance, that the procedure of proposals and rejections will continue for an infinite number of steps. It might also seem that one of the men will be rejected by all the women he courts and ejected from the system. In that case one of the women remains single too. In what follows we shall show that such situations are impossible.

Proof:

(1) The number of men in the community equals the number of women. Therefore, as long as there is a woman in the community with more than one proposal, there is another woman without any proposal.
(2) Once a woman has a proposal, she will always have one, because someone will always be on her waiting list.
(3) When every woman has a proposal, every woman will have exactly one proposal, because the number of men equals the number of women. At this stage the procedure terminates, and there remains only to show that it is indeed possible to get to this stage.
(4) It is possible to get to the stage at which every woman has a proposal, because at every stage the men propose to the women who are next on their list; therefore they cannot backtrack and propose again to the women who rejected them. Because there is a finite number of men and women in the community, and no going back, a stage must be reached at which every woman has a proposal. Based on Step (3) of this proof, the procedure terminates at this stage.

Theorem:

The Gale–Shapley algorithm terminates in a stable matching system.

Proof:

Let us consider a matching system that is an outcome of the Gale–Shapley algorithm. We shall limit our attention to two couples in this system.

Suppose Mr. s prefers Ms. R to his actual mate, Ms. S. It must be shown that Ms. R does not prefer Mr. s to Mr. r, and therefore will refuse to be paired off with him if he suggests it. Indeed, if Mr. s prefers Ms. R, then he must have proposed to her at one of the previous stages. But the fact that he is not paired off with her means that she rejected him. Why was he rejected? Because at that stage Ms. R had a proposal from another man whom she preferred to s (not necessarily r). It is possible that at one of the later stages she rejected this man whom she preferred, in favor of another man whom she preferred even more, etc.

Eventually, Mr. r proposed, and Ms. R preferred him to all the men who had proposed to her up to that stage, including Mr. s. We have now proved that at the end of the Gale–Shapley algorithm there cannot be found a man and a woman who are not paired off with each other but prefer each other to their actual mates. The conclusion is that the algorithm terminates in a stable matching system.

1.9 THE MAXIMUM NUMBER OF COURTSHIP STAGES IN THE GALE–SHAPLEY ALGORITHM

In the previous section we proved that the Gale–Shapley algorithm terminates in a stable matching system after a finite number of steps. We shall now clarify the maximum number of courtship stages in the algorithm, when the number of men equals the number of women.

Example:

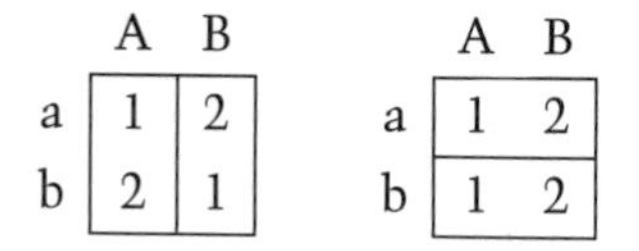

	A	B
a	1	2
b	2	1

	A	B
a	1	2
b	1	2

The courtship procedure of Mr. a and Mr. b:

A	B
a	
b*	
	b

Explanation: Mr. a's and Mr. b's first choice is Ms. A, so both men propose to her at the first stage. Ms. A prefers Mr. a to Mr. b, so she rejects b. At the second stage b proposes to Ms. B and the procedure terminates in the matching system $\begin{pmatrix} A & B \\ | & | \\ a & b \end{pmatrix}$, which is stable.

The courtship procedure in a community of two men and two women may take only one stage – if both men do not propose to the same woman. In this case there are no rejections and the courtship procedure terminates at the first stage.

To sum up: the courtship procedure in a community of two men and two women will end after at most two stages, because at most one man will be rejected at the first stage and he will find a partner at the next stage.

The following examples are drawn from a community of three men and three women.

Example 1

	A	B	C
a	2	3	1
b	3	1	3
c	1	2	2

	A	B	C
a	1	2	3
b	1	3	2
c	2	1	3

A	B	C
a	c	
b*		
	b	

The stable matching system $\begin{pmatrix} A & B & C \\ | & | & | \\ a & c & b \end{pmatrix}$ is obtained after two stages: the first stage at which everyone proposes to his first choice, and the second stage at which Mr. b, who was rejected at the previous stage, proposes to his second choice. The procedure ends at this stage, because previous to this stage Ms. C did not get a proposal, and by now every woman has gotten a proposal.

Example 2

	A	B	C
a	2	3	1
b	3	1	3
c	1	2	2

	A	B	C
a	1	2	3
b	1	3	2
c	3	2	1

A	B	C
a		c
b*		
a		c
		b*
a	b	c

The stable matching system $\begin{pmatrix} A & B & C \\ | & | & | \\ a & b & c \end{pmatrix}$ is obtained after three stages. There are only two rejections in the procedure: Mr. b is rejected twice – at the first stage and at the second stage. Thus, the procedure consists of three stages: the first stage at which everyone proposes to his first choice and b is rejected, the second stage at which b proposes to his second choice and is rejected, and the last stage at which b proposes to his third choice and is not rejected, because previous to this stage Ms. B did not get a proposal, and by now every woman has gotten a proposal.

Example 3

	A	B	C
a	2	3	3
b	3	1	1
c	1	2	2

	A	B	C
a	1	2	3
b	1	2	3
c	2	1	3

A	B	C
a	c	
b*		
a	c*	
	b	
a*	b	
c		
c	b	
	a*	
c	b	a

The stable matching system $\begin{pmatrix} A & B & C \\ | & | & | \\ c & b & a \end{pmatrix}$ is obtained after five stages.

Explanation: Every man is rejected once: at the first stage everyone proposes to his first choice and Mr. b is rejected; at the second stage b proposes to his second choice and c is rejected; at the third stage c proposes to his second choice and Mr. a is rejected; at the fourth stage a proposes to his second choice and is rejected again. At the next stage a proposes to his third choice, Ms. C, who does not reject him, because previous to this stage she did not get a proposal.

We shall show later that five is the maximum number of stages needed in the case of three men and three women.

Before discussing the general case, let us consider an example drawn from a community of four men and four women.

Example:

	A	B	C	D
a	3	2	1	3
b	4	3	2	4
c	1	4	3	2
d	2	1	4	1

	A	B	C	D
a	1	2	3	4
b	1	2	3	4
c	3	1	2	4
d	2	3	1	4

	A	B	C	D
1)	a	c	d	
	b*			
2)	a	c*	d	
		b		
3)	a	b	d*	
			c	
4)	a*	b	c	
	d			
5)	d	b*	c	
		a		
6)	d	a	c*	
			b	
7)	d*	a	b	
	c			
8)	c	a*	b	
		d		
9)	c	d	b*	
			a	
10)	c	d	a	b

Explanation: At the first stage every man proposes to his first choice. Every man is rejected twice, and, after each rejection, every man proposes to his next choice. Thus are added eight more courtship stages. Mr. b is rejected three times and therefore his proposal to his fourth choice adds another stage. Thus the procedure consists of ten stages. In a community of four men and four women, the maximum number of courtship stages in the Gale–Shapley algorithm is ten: the first stage, plus four times two rejections of everyone, plus at most one additional rejection of one man.

Theorem:

If the number of men in the community is n and the number of women is n, then the maximum number of courtship stages is $n^2 - 2n + 2$.

Proof: The algorithm terminates as soon as all the women get a proposal, at which point each woman has exactly one proposal. A maximum number of stages is therefore achieved if at each stage only one man is rejected and if one woman remains without a proposal after all the men have proposed to all the women except for her; that is, they propose to $(n-2)$ women, after the first stage. Since there are n men, that takes $n(n-2)$ stages. Thus, the maximum number of stages can be at most

$$\underset{\text{first stage}}{1} + \underset{\text{rejections}}{n(n-2)} + \underset{\text{last stage}}{1} = n^2 - 2n + 2$$

The previous examples show that this maximum is actually achieved for $n = 3$ and $n = 4$ (calculate!), and similar examples can be constructed for each n.

1.10 GENERALIZATION

I. The number of men does not equal the number of women

In every matching system obtained in a community where the number of men does not equal the number of women, there will be men or, alternatively, women who are not paired off. The Gale–Shapley algorithm and its conclusions can be generalized to this case.

Example:

Women: A, B, C, D, E.

	A	B	C	D	E
a	1	2	1	2	1
b	3	3	2	1	2
c	2	1	3	3	3

Men: a, b, c.

	A	B	C	D	E
a	5	1	2	3	4
b	4	1	2	5	3
c	5	4	1	3	2

(1) The Men Propose:

A	B	C	D	E
	a	c		
	b*			
	a	c*		
		b		
–	a	b	–	c

The procedure terminates in the matching system $\begin{pmatrix} A & B & C & D & E \\ | & | & | & | & | \\ - & a & b & - & c \end{pmatrix}$.

A and D remain single.

(2) The Women Propose:

	a	b	c
	A*	D	B
	C		
	E*		
	C	D*	B
		E	A*
	C	E	B
	D*	A*	
	C	E	B*
A is out.			D
	C*	E	D
	B		
	B	E*	D
		C	
	B	C	D*
			E
D is out.	B	C	E

Here again, the procedure terminates in the matching system

$$\begin{pmatrix} A & B & C & D & E \\ | & | & | & | & | \\ - & a & b & - & c \end{pmatrix}.$$

Exercise:
Check whether the above matching system is stable. Your answer should also take into account the women who are not paired off.

Exercise:
Prove that the Gale–Shapley algorithm always leads to a stable matching system (even when the number of men does not equal the number of women).

II. Existence of a preference list that does not include all members of the opposite sex

In the following example, there are men who would rather be single than be paired off with certain women. Similarly, there are women who would rather be single than be paired off with certain men. In such cases, the preference to stay single rather than be paired off with a certain person is marked with a zero (0). We shall now show that the Gale–Shapley algorithm can be generalized to this case too.

Example:
The preference list is:

Women: A, B, C.

	A	B	C
a	1	1	0
b	2	2	0
c	0	3	0
d	3	4	1

Men: a, b, c, d.

	A	B	C
a	3	1	2
b	0	1	2
c	1	2	3
d	1	0	2

The Men Propose:

	A	B	C
	c*	a	
	d	b*	
	d	a	
b and c are out.		c*	b*
C is out.	d	a	c*

The matching system obtained is $\begin{pmatrix} A & B & C \\ | & | & | \\ d & a & - \end{pmatrix}$.

Show that even those who are left out cannot undermine the system.

III. Possible Indifference

Until now we have been looking at cases where every community member has had a strict preference for members of the opposite sex. In other words, no community member has been indifferent to a choice between two or more members of the opposite sex. As we said in Section 1.2, the assumption that there is no indifference was introduced in order to simplify the procedure. We shall now see that this requirement can be dropped.

Let us see what happens when indifference is allowed. A community member who is indifferent to a choice between two or more members of the opposite sex and yet obliged to rank them in order of preference might say: "My first choice is Ms. B; as for my second choice, I am indifferent between A and D; as for my third choice, I am indifferent between C and E; and my fourth choice is F."

It turns out that there can be a stable matching system even when there is indifference.

Definition:

A matching system is called *stable* if under it there cannot be found a man and a woman who are not paired off with each other but prefer each other to their actual mates.

Remark:

It follows from this definition that we assume that a man will not leave his mate for another woman when he is indifferent between the two, and a woman will not leave her mate for another man when she is indifferent between the two.

Up to now we have made the assumption that the preference order of each member contains no indifference. We can easily dispense with this constraint by assigning an arbitrary strict preference whenever there is indifference.

Consider the following example:

Women: A, B, C. **Men:** a, b, c.

	A	B	C
a	3	1	3
b	2	2	1
c	1	3	2

	A	B	C
a	1	2	2
b	1	1	2
c	1	2	3

In this example, Mr. a's first choice is Ms. A, but he is indifferent to a choice between B and C, who occupy a lower rank in his order of preferences. Mr. b's first choice is A and B, but he is indifferent to a choice between them; his second choice is Ms. C. There is no indifference in the women's preferences.

Let us try to follow the Gale–Shapley algorithm.

A	B	C
a		
c		

↖b↗

According to the preference structure, a and c propose to A, who is their first choice, but b hesitates, because he prefers A and B equally. To proceed any further, we need to arbitrarily change the given preference structure to a preference structure in which there is no indifference. Specifically, we need to let strict preferences stand and replace indifference, wherever it occurs, by a strict preference.

For example:

	A	B	C
a	3	1	3
b	2	2	1
c	1	3	2

	A	B	C
a	1	3	2
b	2	1	3
c	1	2	3

Question: Why did we write 3 in b's row of preferences?

Given this preference structure, we can use the Gale–Shapley algorithm to obtain a stable matching system.

In the male courtship procedure:

A	B	C
a*	b	
c		
c	b	a

The stable matching system is $\begin{pmatrix} A & B & C \\ | & | & | \\ c & b & a \end{pmatrix}$.

Is this matching system stable in the original preference structure too?

Does the relation indicated below threaten the system's stability?

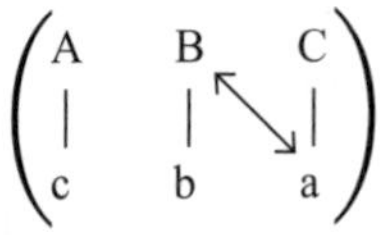

$$\begin{pmatrix} A & B & & C \\ | & | & \nwarrow\!\!\searrow & | \\ c & b & & a \end{pmatrix}$$

Indeed, B prefers a to her actual mate, but a does not prefer B to C; rather, he is indifferent to a choice between them. An examination of all the possible relations reveals that deviation from the matchmaker's recommendation is impossible. Thus, the system is stable in the original preference structure too.

We shall now use another preference structure in which there is no indifference. Here again, we shall let the original strict preferences stand and replace indifference, wherever it occurs, by a strict preference. For example:

	A	B	C
a	3	1	3
b	2	2	1
c	1	3	2

	A	B	C
a	1	2	3
b	2	1	3
c	1	2	3

We shall have the men propose according to the Gale–Shapley algorithm.

	A	B	C
	a*	b	
1)	c		
	c	b*	
2)		a	
	c	a	
3)	b*		
	c	a	b

The stable matching system is $\begin{pmatrix} A & B & C \\ | & | & | \\ c & a & b \end{pmatrix}$.

This matching system is stable in the original preference structure too (verify it!).

Summary: To obtain a stable matching system, given a preference structure in which there is indifference, one constructs an alternative preference structure in which there is no indifference and finds a stable matching system using the Gale–Shapley algorithm. This matching system will be stable in the original preference structure too. Note that the Gale–Shapley algorithm, which leads to one stable matching system when there is no indifference, may lead to several stable matching systems when there is indifference.

Claim: *Every stable matching system in a "revised" preference structure in which there is no indifference is also stable in the original preference structure in which there is indifference.*

Proof: Let us assume, on the contrary, that the stable matching system in the revised preference structure is unstable in the original preference structure in which there is indifference. Now, according to the original preference structure, there exist Ms. X and Mr. y who are not paired off with each other but prefer each other to their actual mates. Because preference relations (in contrast to indifference relations) do not change in the conversion to the revised preference structure in which there is indifference, X and y prefer each other to their actual mates in this preference structure too. Therefore, this matching system is unstable in the revised preference structure, which contradicts our assumption.

The contradiction proves that the assumption made at the beginning of the proof is incorrect; therefore, the stable matching in the revised preference structure is also stable in the original preference structure where indifference occurs.

I.II EXERCISES

1. Given the following preference structure, find a stable matching system using the Gale–Shapley algorithm when the men propose:

Women: A, B, C.

	A	B	C
a	1	1	3
b	3	2	1
c	2	3	2
d	4	4	4

Men: a, b, c, d.

	A	B	C
a	3	2	1
b	1	2	3
c	2	1	3
d	1	2	3

2. Given the following preference structure, find a stable matching system using the Gale–Shapley algorithm (1) when the men propose, and (2) when the women propose:

Women: A, B, C.

	A	B	C	D	E
a	3	1	1	1	2
b	2	2	0	2	1
c	1	3	2	0	0

Men: a, b, c, d.

	A	B	C	D	E
a	1	3	2	5	4
b	1	2	0	3	4
c	3	2	1	0	0

3. Given the following preference structure, find a stable matching system using the Gale–Shapley algorithm, when the women propose:

Women: A, B, C.

	A	B	C
a	1	1	0
b	2	0	1
c	3	2	3
d	4	0	2

Men: a, b, c, d.

	A	B	C
a	0	0	1
b	3	1	2
c	2	1	0
d	2	3	1

4. A community consists of three women and three men. According to the following preference structure, one man and one woman in the community are indifferent to a choice between members of the opposite sex.

Women: A, B, C.

	A	B	C
a	1	2	1
b	3	3	1
c	2	1	1

Men: a, b, c.

	A	B	C
a	2	1	3
b	1	1	1
c	1	2	3

Determine whether each of the following matching systems is stable. If a system is unstable, find the couple(s) responsible for the instability. If a system is stable, explain why.

i.

$$\begin{pmatrix} A & B & C \\ | & | & | \\ c & b & a \end{pmatrix}$$

ii.

$$\begin{pmatrix} A & B & C \\ | & | & | \\ c & a & b \end{pmatrix}$$

iii.

$$\begin{pmatrix} A & B & C \\ | & | & | \\ b & c & a \end{pmatrix}$$

iv.

$$\begin{pmatrix} A & B & C \\ | & | & | \\ b & a & c \end{pmatrix}$$

v.

$$\begin{pmatrix} A & B & C \\ | & | & | \\ a & b & c \end{pmatrix}$$

vi.

$$\begin{pmatrix} A & B & C \\ | & | & | \\ a & c & b \end{pmatrix}$$

Are there other possible matching systems in this community? Support your answer.

5. A community consists of three women and three men. According to the following preference structure, one of the men is indifferent to a choice between all the women in the community:

Women: A, B, C. **Men:** a, b, c.

Women:

	A	B	C
a	1	1	3
b	2	3	1
c	3	2	2

Men:

	A	B	C
a	2	3	1
b	1	1	1
c	1	2	3

(1) Find a stable matching system using the Gale–Shapley algorithm.
(2) Find another stable matching system.

6. In the following preference structure Ms. B is indifferent to a choice between all the men in the community. Find a stable matching system for this preference structure when the men propose, assuming B's preferences are the following:

(1) B prefers a, d, b, c, in that order.
(2) B prefers d, c, b, a, in that order.

Is the same matching system obtained in both cases?

Women: A, B, C. **Men:** a, b, c, d.

Women:

	A	B	C
a	1	1	2
b	2	1	1
c	0	1	3
d	0	1	0

Men:

	A	B	C
a	3	1	2
b	1	2	0
c	2	1	0
d	0	1	0

7. In the following preference structure Mr. a is indifferent to a choice between Ms. A and Ms. B. Find a stable matching system in the following two cases:

(1) a prefers A, B, C, D, in that order.
(2) a prefers B, A, C, D, in that order.

Check whether the same matching system is obtained in both cases, using the Gale–Shapley algorithm when the men propose to the women and vice versa.

Women: A, B, C, D.

	A	B	C	D
a	0	3	1	1
b	3	4	2	0
c	1	0	3	3
d	0	1	4	2
e	2	2	5	0

Men: a, b, c, d, e.

	A	B	C	D
a	1	1	2	3
b	1	0	2	3
c	2	3	1	4
d	1	2	0	3
e	0	1	2	0

8. Find all stable matching systems obtained by the Gale–Shapley algorithm when the men propose.

Women: A, B, C.

	A	B	C
a	2	2	3
b	1	1	1
c	1	2	2

Men: a, b, c.

	A	B	C
a	1	1	2
b	2	3	1
c	2	1	2

9. Given the following preference structure, find a stable matching system:

Women: A, B, C, D.

	A	B	C	D
a	0	2	1	4
b	1	3	0	1
c	1	1	2	2
d	2	2	3	3

Men: a, b, c, d.

	A	B	C	D
a	1	1	2	0
b	3	1	2	2
c	1	2	3	4
d	0	1	2	0

1.12 THE GALE–SHAPLEY ALGORITHM AND THE ASSIGNMENT PROBLEM

The matching problem discussed in the preceding sections, while interesting, is of limited practical value. Real-life courtship behavior bears little resemblance to the Gale–Shapley algorithm. For example, in a real-life matching of n men and n women, the matches are made haphazardly, not procedurally.

In this section we shall show that, by a slight generalization, the Gale–Shapley algorithm can be of great value also in everyday life. Indeed, the medical school admissions problem to which we now turn our attention is only one example among many demonstrating that mathematics can be used to solve real-life problems.

The medical school admissions problem

In many countries a large number of applicants seek admission to a small number of medical schools. The competition for this limited number of places poses several problems. Because the number of candidates is greater than the medical school admissions quotas, many candidates apply to several medical schools. Imagine a situation in which the number of applications submitted to the admissions office of a certain medical school exceeds its admissions quota. In such a case the admissions office must evaluate all applicants and decide which ones to accept and which ones to reject. The medical school will offer admission to some applicants if there are vacancies, and refuse admission to other (less-qualified) applicants even if there are vacancies.

A priori, the solution seems obvious: the medical school should accept the most-qualified applicants until the desired quota is filled. This solution fails the reality test, however, because one cannot assume that all who are offered admission will not accept what in their opinion are better offers from other medical schools. If they do choose other medical schools, then the admissions quota of the medical school in question will not be filled.

Therefore, we may propose another solution: the medical school, to fill its desired quota, needs to offer admission to a number of applicants that is greater than that quota. But this solution is also unsatisfactory, because if not enough students decline the offer the number of acceptances might exceed the absorption capacity of the institution.

Neither of these two outcomes can be improved, because the admissions office lacks relevant data about the applicants:

(1) It is not known whether a given applicant has also applied elsewhere.
(2) It is not known how each applicant ranks the medical schools to which he has applied.
(3) It is not known which of the other medical schools will offer to admit each applicant.

As a result of all this uncertainty, medical schools must prepare for the possibility that the entering class will fail to meet the quota and that only a portion of the students admitted will indeed be outstanding students.

The usual admissions procedure poses problems for the applicants as well:

(1) Not without good cause do applicants become suspicious when asked to declare their order of preferences. For example, an applicant who is asked to list in his application all other medical schools applied to in order of preference may feel that by telling a medical school it is, say, his third choice he will be hurting his chances of being admitted.
(2) Assuming that medical schools draw up waiting lists, an applicant may be informed that he is not admitted but may be admitted later if a vacancy occurs. This situation poses new problems. Suppose the applicant is accepted by one medical school and placed on the waiting list of another that he prefers.
    1. Should he play safe by accepting the first or take a chance that the second will admit him later?

2. Is it ethical to accept one school without informing the other and then withdraw this acceptance because the second school admits him later?

The source of the difficulties here described is the lack of data in the admissions offices of the medical schools on the one hand, and the (not wholly unfounded) suspicions of the applicants on the other.

One may ask how all this uncertainty relates to the matching problem. As we shall see, a generalization of the Gale–Shapley algorithm yields a stable solution, which removes all the above-mentioned difficulties.

It is possible to solve these problems and assign applicants among medical schools by means of an independent "placement center," where data is collected and assignments[6] are made in the following manner:

(i) Each applicant turns to the placement center and submits a list of the medical schools that he is willing to accept, omitting only those medical schools that he is unwilling to accept even if there are vacancies.

(ii) Each applicant ranks the medical schools that he is willing to accept in order of preference. (For convenience, we assume that there are no instances of indifference; that is, there is a strict order of preferences.) He then submits his ranking to the placement center.

(iii) The placement center sends each medical school a list of all applicants who applied to that medical school.

(iv) Each medical school announces its admissions quota.

(v) Each medical school submits to the placement center a list of all applicants ranked in accordance with its own preferences and informs the placement center which applicants will not be admitted even if there are vacancies.

[6] *Assignment* is the association of applicants and places. The problem considered in this section is a special case of a more complicated assignment problem.

Now, the placement center has all the necessary information to revise each applicant's preference order by omitting all the medical schools that will never admit him. At this point, the procedure for the assignment of applicants to medical schools can start by implementing the Gale–Shapley algorithm as follows:

(a) All applicants are referred to the medical school of their first choice according to the new preference order. The medical schools whose number of applicants at this stage is less than or equal to their quota remain on a waiting list. If the number of candidates is greater than the quota, the applicants are selected according to the preference order of the medical school until the quota is filled, and any applicants left are rejected.

(b) The applicants who have been rejected are then referred to their next-choice medical school. They are added to the list from the previous stage. If the number is less than or equal to the quota, all of them remain on the new waiting list. Otherwise, they are selected according to the preference order of the medical school until the quota is filled, and any applicants left are rejected.

The procedure continues, until there are no more rejections. The procedure terminates when every applicant either is on a waiting list or has been rejected by every medical school to which he is willing and able to apply. At this point all the medical schools admit everyone on their waiting list. The rest will not go to any medical school. The assignment thus achieved is stable.

1.13 EXERCISES

1. Given the following preference structure, find a stable assignment of applicants to medical schools, using the Gale–Shapley algorithm.

Medical schools:

A – quota of 3

B – quota of 4

C – quota of 6

Applicants: a, b, c, d, e, f, g, h, i, j, k, l, m, n, o.

The preference structure:

	a	b	c	d	e	f	g	h	i	j	k	l	m	n	o
A	1	5	2	4	3	10	6	9	11	7	8	14	12	13	15
B	1	4	2	3	6	14	5	13	12	11	7	8	9	15	10
C	5	2	1	14	6	12	13	7	8	9	10	11	15	3	4

	a	b	c	d	e	f	g	h	i	j	k	l	m	n	o
A	1	3	1	1	2	2	3	2	2	3	1	1	2	2	1
B	2	1	3	2	1	3	2	1	3	1	2	3	1	3	2
C	3	2	2	3	3	1	1	3	1	2	3	2	3	1	3

2. Given five medical schools:

A – quota of 9

B – quota of 6

C – quota of 7

D – quota of 5

E – quota of 4

and twenty applicants: a, b, c, d, e, f, g, h, i, j, k, l, m, n, o, p, q, r, s, t.

Their preference structure is:

	a	b	c	d	e	f	g	h	i	j	k	l	m	n	o	p	q	r	s	t
A	8	9	1	0	0	0	2	3	0	6	4	5	0	0	10	0	0	0	0	7
B	1	2	3	4	5	6	7	8	0	0	0	0	0	0	0	0	0	0	0	0
C	1	2	0	0	0	0	0	3	6	0	4	5	8	9	0	0	0	0	0	7
D	0	0	0	4	0	0	0	0	0	0	0	0	0	0	0	1	2	3	5	0
E	2	3	0	4	1	0	5	0	0	0	0	0	0	0	0	0	0	0	0	0

	a	b	c	d	e	f	g	h	i	j	k	l	m	n	o	p	q	r	s	t
A	1	0	2	1	2	1	5	1	0	2	1	1	0	0	0	0	0	4	1	1
B	2	0	3	0	1	0	1	2	0	0	2	2	4	0	0	0	1	0	2	0
C	3	0	4	0	3	0	2	3	0	0	3	3	1	1	1	0	0	1	3	2
D	0	1	5	0	0	2	3	4	0	0	4	0	2	2	0	1	0	2	4	0
E	0	2	1	0	0	3	4	0	1	1	5	0	3	0	0	0	0	3	5	0

Find a stable assignment when medical schools "make bids" to applicants, using the Gale–Shapley algorithm.

3. Given five medical schools:

A – quota of 3
B – quota of 2
C – quota of 1
D – quota of 3
E – quota of 1

and ten applicants: a, b, c, d, e, f, g, h, i, j.

Their preference structure is:

	a	b	c	d	e	f	g	h	i	j
A	10	5	1	4	2	7	3	6	8	9
B	10	5	1	4	2	7	3	6	8	9
C	10	5	1	4	2	7	3	6	8	9
D	10	5	1	4	2	7	3	6	8	9
E	10	5	1	4	2	7	3	6	8	9

	a	b	c	d	e	f	g	h	i	j
A	1	1	1	1	1	1	1	1	1	1
B	2	2	2	2	2	2	2	2	2	2
C	3	3	3	3	3	3	3	3	3	3
D	4	4	4	4	4	4	4	4	4	4
E	5	5	5	5	5	5	5	5	5	5

Find a stable assignment, using the Gale–Shapley algorithm:

(1) when applicants apply to medical schools.
(2) when medical schools make bids to applicants.

What is interesting about these two assignments?

4. Prove that applying the Gale–Shapley algorithm to the assignment problem described in the previous section yields a stable assignment.

5. Describe how the Gale–Shapley algorithm can be used to solve the assignment problem when medical schools make bids to applicants.

1.14 OPTIMALITY

It has been seen that some preference structures yield more than one stable matching system. This raises a few questions.

(1) Is there one stable matching system that is everyone's favorite? Assuming that there is no indifference, the answer is no, because if there are two stable matching systems, then at least one man is paired off with a different woman in the second system, and necessarily prefers one system to the other.
(2) Is there one stable matching system that is the men's favorite? Surprisingly, the answer is yes. The same goes for the women: there is one stable matching system that is the women's favorite.

We shall illustrate each of these points by way of example.

Example:

Given the following preference structure:

Women: A, B, C, D.

	A	B	C	D
a	3	4	1	1
b	2	2	3	4
c	4	1	2	3
d	1	3	4	2

Men: a, b, c, d.

	A	B	C	D
a	2	1	4	3
b	3	2	1	4
c	2	4	3	1
d	4	2	1	3

When the men propose, the following matching system is obtained by the Gale–Shapley algorithm (verify it!):

System 1:

$$\begin{pmatrix} A & B & C & D \\ | & | & | & | \\ a & d & b & c \end{pmatrix}$$

When the women propose, the following matching system is obtained by the Gale–Shapley algorithm (verify it!):

System 2:

$$\begin{pmatrix} A & B & C & D \\ | & | & | & | \\ d & b & c & a \end{pmatrix}$$

The following matching system, which is not obtained by the Gale–Shapley algorithm, is stable too (verify it!):

System 3:

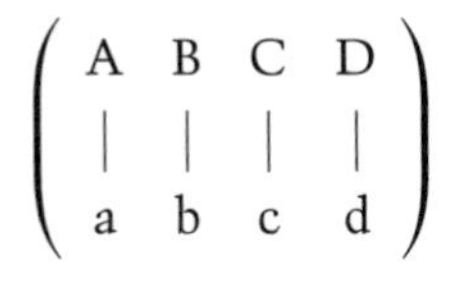

There are no other stable matching systems for this preference structure (verify it!).

In the following table, we include what preference ranking is assigned to each man and woman in the above three stable systems.

	Male Courtship	**Female Courtship**	**Other System**
A B C D:	3 3 3 3	1 2 2 1	3 2 2 2
a b c d:	2 1 1 2	3 2 3 4	2 2 3 3
	System 1	**System 2**	**System 3**

Let us see which system each woman will prefer and which system each man will prefer:

Ms. A will prefer System 2, because she gets her first choice there.
Ms. B will prefer System 2 or 3, because she gets her second choice in both and her first choice in neither.
Ms. C will prefer System 2 or 3 for the same reason that Ms. B prefers them.
Ms. D will prefer System 2, because she gets her first choice there.

Note: Systems 2 and 3, which are best for Ms. B and Ms. C, are equally good for them. But only System 2 is best for all the women.

A similar situation obtains among the men: in System 1, obtained by the Gale–Shapley algorithm when the men propose, each man gets the best choice that he can get in any stable matching system.

Again, it is worth noting that for Mr. a, System 3 is equally good. The stress is on the word "equally"; that is, System 3 is as good as, but not better than, System 1. Only System 1 is best for all the men.

To sum up: the matching system that is obtained when the men propose is *optimal for every man* and the matching system that is obtained when the women propose is *optimal for every woman*, when "optimal" is defined as follows:

Definition:
A stable matching system is called *optimal for a given man* if he is at least as well off under it as under any other stable matching system. Similarly, a stable matching system is called *optimal for a given woman* if she is at least as well off under it as under any other stable matching system.

Note: We are comparing only stable matching systems. The system in question must be stable, and bears comparison only to other stable matching systems. The satisfaction of a single individual with an unstable matching system is irrelevant, because an unstable matching system will not last and will be undermined by internal deviations.

Optimality Theorem:
For every preference structure, the matching system obtained by the Gale–Shapley algorithm, when the men propose, is optimal for the men. The matching system obtained by the Gale–Shapley algorithm, when the women propose, is optimal for the women.

Preliminaries to the Proof of the Theorem:
For the sake of the proof, we define the following terms:

Definition:
Ms. K is called *possible* for Mr. k if there exists a stable matching system in which they are paired off with each other.

Definition:
Ms. K is called *impossible* for Mr. k if there does *not* exist a stable matching system in which they are paired off with each other.

Note: To prove that Ms. K is possible for Mr. k, it is enough to indicate one stable matching system in which they are paired off with each other. However, to prove that Ms. K is impossible for Mr. k, it must be shown that there is no stable matching system in which they are paired off with each other.

Proof of the Theorem:
We shall prove the theorem for the case where the number of men and women is equal, there is no indifference, and no community member prefers staying single. We shall examine the procedure when the men propose. The proof for the case where the women propose is obtained by reversing the gender roles.

The theorem is proved if it is seen that every man rejected by a woman is rejected only by a woman who is impossible for him; that is, every man is rejected only by women with whom he cannot be paired off in any case in a stable matching system.

If that is true, then every man, after proposing according to his preferences, will eventually be paired off with the woman he prefers most among the women possible for him, that is, with the woman he prefers most among all the stable systems.

Claim:
In any male courtship procedure, if Mr. x is rejected by Ms. X, then she is impossible for him.

Proof of the Claim:
Let us examine the first stage of the Gale–Shapley algorithm. All the men propose to their first choice. Every woman who gets more than one proposal selects her favorite from among those who have proposed to her and informs him that he is on her waiting list, but makes him no promises. She informs the others that they have been rejected. Women who get no proposals at this stage wait for the next stage.

We shall consider Ms. X, who at the first stage gets proposals from Mr. x and Mr. y, and maybe others. Suppose she prefers x to y and so rejects y. Then suppose there is an alternative matching system in which Mr. y is paired off with Ms. X. This is represented below.

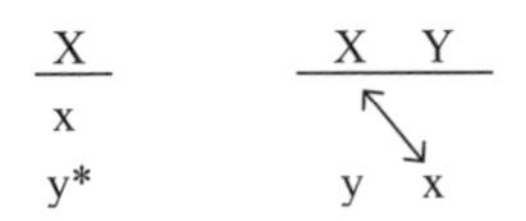

Let us examine the alternative matching system. X prefers x to y; we know that at the first stage of the courtship procedure X rejected y in preference to x. For his part, x prefers X to Y; indeed, X is his first choice because he proposed to her already at the first stage of the courtship procedure. Therefore, the alternative matching system cannot be stable.

Thus we have proved that a man who is rejected at the first stage is rejected by a woman who is impossible for him. Now we shall prove that a man who is rejected at any stage of the procedure is rejected by a woman who is impossible for him.

Suppose the contrary is true. That is, suppose there exists a stage (of course, as we proved, it cannot be the first stage) at which a man is rejected by a woman who is possible for him. Thus, there is a *first time* that this happens. That is, up to this stage, all the women who rejected men were impossible for those men and now, for the first time, there comes along Ms. Z who rejects Mr. z because she prefers Mr. w to him. But Z is possible for z; that is, there exists an alternative stable matching system in which they are paired off, as shown below.

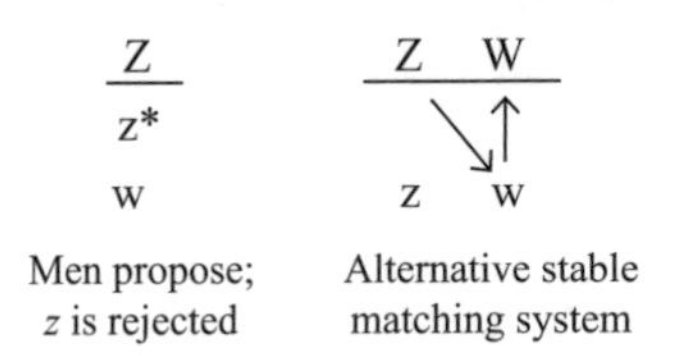

Men propose; *z* is rejected

Alternative stable matching system

No doubt, Ms. Z prefers w to z, because she rejected z during the courtship procedure. Because the alternative system is assumed to be stable, w prefers W to Z. (Explain why!)

If w prefers W to Z, then during the courtship procedure he proposed to W before he proposed to Z, and if he then gets to Z, then he was rejected by W. Hence w was already rejected at a previous stage by a woman who was possible for him and therefore this is not the first time that such a rejection has occurred, contrary to our assumption that it is.

We have come up against a contradiction; the contradiction proves that our assumption is wrong. The assumption was that there exists a case where a woman rejects a man in spite of the fact that she is possible for him. Thus, such a case does not exist. Therefore, we have proved that every man who is rejected by a woman is rejected by a woman who is impossible for him.

As we said, this conclusion completes the proof of the Optimality Theorem, according to which the matching system obtained by the Gale–Shapley algorithm, when the men propose, is optimal for the men. The proof of the second part of the Optimality Theorem is similar.

We see, then, that when the men take the initiative to propose and the women remain passive, the result is favorable to the men: they obtain the best choices for themselves from among all the stable systems. This result holds when the number of men is not equal to the number of women, as well as for assignment problems of a more general nature, such as the one described in Section 1.12 (p. 37).

In all these cases, the side that proposes obtains the most favorable result from among all the stable results.

This is not the case when there is indifference. Let us examine the example in Section 1.10, III (p. 30). Here there is indifference in the original preference structure.

Women: A, B, C.

	A	B	C
a	3	1	3
b	2	2	1
c	1	3	2

Men: a, b, c.

	A	B	C
a	1	2	2
b	1	1	2
c	1	2	3

To get a stable matching system, we have arbitrarily changed the original preference structure to a preference structure in which there is no indifference. We present two possibilities:

i.

	A	B	C
a	3	1	3
b	2	2	1
c	1	3	2

	A	B	C
a	1	3	2
b	2	1	3
c	1	2	3

ii.

	A	B	C
a	3	1	3
b	2	2	1
c	1	3	2

	A	B	C
a	1	2	3
b	2	1	3
c	1	2	3

If we implement the Gale–Shapley algorithm when the men propose, we get two different stable matching systems:

$\begin{pmatrix} A & B & C \\ | & | & | \\ c & b & a \end{pmatrix}$, for the first preference structure, and $\begin{pmatrix} A & B & C \\ | & | & | \\ c & a & b \end{pmatrix}$, for the second.

Each of these systems is optimal for the revised preference structure in which there is no indifference, but it cannot be said that they are both optimal for the original structure. In the above example, the two matching systems are equally good for Mr. a, because he is indifferent to the choice between B and C, but the first system is better for Mr. b, because he prefers B to C. Note that each matching system is optimal for the revised preference structure in which there is no indifference, but not necessarily for the original preference structure in which there is indifference.

1.15 EXERCISES

1. At the beginning of Section 1.14 (p. 43) we discussed a case where there are two optimal matching systems for one man. Can a

situation exist where there are two optimal matching systems for all the men?

2. Consider the example at the beginning of Section 1.14.

(1) Are all the women possible for Mr. d? If not, which women are impossible?
(2) Which women are possible for Mr. a?
(3) Which women are possible for more than one man? Which men are they possible for?

3. Given the following preference structure:

Women: A, B, C.

	A	B	C
a	1	1	2
b	2	3	3
c	3	2	1

Men: a, b, c.

	A	B	C
a	2	3	1
b	3	2	1
c	1	2	3

(1) Is A possible for a?
(2) Is A possible for b?
(3) Is B possible for b?

4. Given the following preference structure:

Women: A, B, C, D, E.

	A	B	C	D	E
a	1	4	4	1	3
b	4	5	2	2	4
c	2	2	3	3	1
d	3	3	1	4	5
e	5	1	5	5	2

Men: a, b, c, d, e.

	A	B	C	D	E
a	3	4	1	2	5
b	5	1	4	2	3
c	3	4	5	1	2
d	3	2	4	5	1
e	2	3	1	4	5

(1) Find an optimal matching system for the men.
(2) Find an optimal matching system for the women.
(3) Is the same matching system obtained in both cases?

5. Given the following preference structure:

Women: A, B, C. **Men:** a, b, c, d.

	A	B	C
a	1	1	4
b	3	3	1
c	2	4	2
d	4	2	3

	A	B	C
a	2	3	1
b	2	1	3
c	3	1	2
d	1	3	2

(1) Find an optimal matching system for the women.

(2) Find an optimal matching system for the men.

(3) Is the same matching system obtained in both cases?

6. Given the following preference structure:

Women: A, B, C. **Men:** a, b, c.

	A	B	C
a	3	2	1
b	2	1	2
c	1	3	3

	A	B	C
a	1	1	2
b	1	2	2
c	1	2	3

Find all the stable matching systems obtained by the Gale–Shapley algorithm when the men propose. Are they optimal for the men in the original preference structure too?

7. Given the following preference structure:

Women: A, B, C, D. **Men:** a, b, c, d.

	A	B	C	D
a	4	4	3	1
b	3	3	4	2
c	2	1	1	3
d	1	2	2	4

	A	B	C	D
a	1	1	2	2
b	1	2	3	4
c	2	1	4	3
d	2	3	1	4

(1) Find all the stable systems obtained by the Gale–Shapley algorithm when the men propose to the women and vice versa.

(2) What can be concluded from these findings? (This will be further discussed in the next section.)

1.16 CONDITION FOR THE EXISTENCE OF A UNIQUE STABLE MATCHING SYSTEM

In previous sections we looked at preference structures with more than one stable matching system.

The male courtship procedure leads to a matching system at one extreme, which favors the interests of the men. Similarly, the female courtship procedure leads to a matching system at the other extreme, which favors the interests of the women.

We have seen that, in general, there exist several stable matchings for a given preference structure. We shall now discuss cases where a unique stable matching exists for a given preference structure.

Example:

Consider the preference structure discussed in Section 1.6, Example 1 (p. 16):

	A	B	C	D
a	3	3	2	3
b	4	1	3	2
c	2	4	4	1
d	1	2	1	4

	A	B	C	D
a	1	2	3	4
b	1	4	3	2
c	2	1	3	4
d	4	2	3	1

We found that when the men propose, the matching system obtained is:

$$\begin{pmatrix} A & B & C & D \\ | & | & | & | \\ c & d & a & b \end{pmatrix}.$$

Now we shall apply the Gale–Shapley algorithm when the women propose:

Stage 1:

a	b	c	d
	B	D	A*
			C

Stage 2:

a	b	c	d
	B	D*	C
		A	

Stage 3:

a	b	c	d
	B*	A	C
	D		

Stage 4:

a	b	c	d
	D	A	C*
			B

Stage 5:

a	b	c	d
C	D	A	B

At this stage the algorithm terminates and the matching system obtained is:

$$\begin{pmatrix} A & B & C & D \\ | & | & | & | \\ c & d & a & b \end{pmatrix}.$$

This is the same matching system that was obtained when the men proposed.

Are there other stable matching systems?

It seems that it will be necessary to check twenty-three additional systems, but in fact this will not be necessary. There is a unique stable matching system for the given preference structure. This is derived from the following theorem:

Theorem:

Assuming that there is no indifference, if the male courtship procedure and the female courtship procedure lead to the same matching system, then there is a unique stable matching system for the given preference structure.

Proof: Let 1 denote the matching system obtained through the male courtship procedure (the same system, as we said, that was obtained through the female courtship procedure). Let 2 be a stable matching system different from 1. We shall prove that System 2 cannot be stable. System 2 is different because it has a couple (a, B) that does not appear in System 1. In System 1, a is paired off with A and B with b.

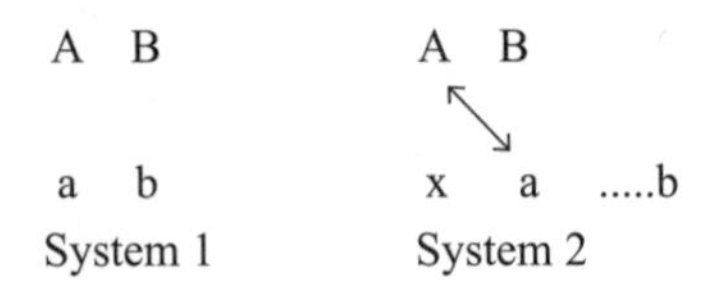

Let x denote A's mate in System 2, where x and a are not the same person.

No doubt, a prefers A to B, because A is the optimal choice for him and this choice is different from B, and because indifference was assumed not to exist.

Similarly, A will prefer a to x, because a is the optimal choice for her and this choice is different from x. It follows that this system is unstable.

I.17 EXERCISES

1. Is $\begin{pmatrix} A & B & C & D & E \\ | & | & | & | & | \\ d & c & e & a & b \end{pmatrix}$ a unique stable matching system for the following preference structure?

Women: A, B, C, D, E. **Men:** a, b, c, d, e.

	A	B	C	D	E
a	5	4	5	4	5
b	4	5	4	5	3
c	1	1	3	3	1
d	2	3	1	1	2
e	3	2	2	2	4

	A	B	C	D	E
a	1	5	2	4	3
b	4	1	2	5	3
c	5	1	2	3	4
d	1	2	3	5	4
e	2	4	1	3	5

2. Is there a common optimal matching system for men and women for the following preference structure?

Women: A, B, C, D. **Men:** a, b, c, d, e.

	A	B	C	D
a	0	1	1	5
b	1	5	3	1
c	0	2	2	4
d	3	4	0	2
e	2	3	4	3

	A	B	C	D
a	1	3	0	2
b	3	1	2	4
c	4	1	2	3
d	3	2	1	0
e	2	3	1	4

3. Is there a unique stable matching system for the following preference structure?

Women: A, B, C. **Men:** a, b, c, d.

	A	B	C
a	3	3	1
b	1	1	2
c	2	4	0
d	4	2	3

	A	B	C
a	1	2	3
b	2	3	1
c	2	1	0
d	1	3	2

4. Is $\begin{pmatrix} A & B & C & - \\ | & | & | & | \\ d & c & a & b \end{pmatrix}$ a unique stable matching system for the following preference structure? If not, find another stable matching system.

Women: A, B, C. **Men:** a, b, c, d.

	A	B	C
a	1	4	3
b	2	3	4
c	0	2	2
d	3	1	1

	A	B	C
a	2	3	1
b	0	2	1
c	1	2	3
d	1	3	2

1.18 DISCUSSION

Gale and Shapley were the first to ask whether their algorithm for matching men and women was applicable to the college admissions problem. As they wrote in the conclusion of their paper, "In making the special assumptions needed in order to analyze our problem mathematically, we necessarily moved further away from the original college admissions question, and eventually in discussing the marriage problem, we abandoned reality altogether and entered the world of mathematical make-believe. ... It was our opinion, too, however, that some of the ideas introduced here might usefully be applied to certain phases of the admissions problem."

What Gale and Shapley did not know at the time was that the Association of American Medical Colleges had already for ten years been applying the Gale–Shapley algorithm to the task of assigning interns to hospitals in the United States. By a process of trial and error that spanned over half a century, the Association in 1951 adopted the procedure, later rediscovered by Gale and Shapley, that was hospital-optimal. A detailed description of this procedure can be found in *Two-Sided Matching* by A. E. Roth and M. Sotomayor.[7] The book contains many extensions that evolved after the appearance of the paper by Gale and Shapley.

1.19 REVIEW EXERCISES

1. Given the preference structure:

	A	B	C
a	1	2	4
b	2	1	1
c	3	3	2
d	4	4	3

	A	B	C
a	1	2	3
b	1	2	3
c	2	3	1
d	3	1	2

Is $\begin{pmatrix} A & B & C & - \\ | & | & | & | \\ a & c & d & b \end{pmatrix}$ a stable matching system? Explain.

[7] Roth, A. E. and Sotomayor, M. 1990. *Two-sided matching: a study in game-theoretic modeling and analysis*. Cambridge: Cambridge University Press.

2. Given the preference structure:

	A	B	C	D
a	2	4	2	3
b	1	1	3	2
c	3	2	4	4
d	4	3	1	1

	A	B	C	D
a	2	1	3	4
b	3	1	2	4
c	3	2	4	1
d	1	2	4	3

(1) Find a stable matching system using the Gale–Shapley algorithm when the men propose.

(2) Is there a unique stable matching system for this preference structure? Explain.

3. Given the preference structure:

	A	B	C	D	E
a	1	1	2	3	3
b	2	3	1	1	2
c	3	2	3	2	1

	A	B	C	D	E
a	2	1	3	4	5
b	3	1	2	5	4
c	3	1	4	2	5

(1) Find a stable matching system using the Gale–Shapley algorithm when the women propose.

(2) Which women do not have a mate? Will these women remain without a mate in every other stable matching? Support your answer.

4. Given the preference structure:

	A	B	C	D
a	3	0	3	4
b	4	1	1	3
c	1	0	2	1
d	2	2	4	2

	A	B	C	D
a	1	2	0	0
b	1	3	2	4
c	0	1	2	3
d	3	1	2	4

Find a stable matching system.

5. Given the preference structure:

	A	B	C
a	3	2	1
b	1	1	2
c	2	1	3

	A	B	C
a	1	1	2
b	2	1	3
c	3	2	1

(1) Find a stable matching system, using the Gale–Shapley algorithm when the men propose to the women and vice versa.

(2) Do these two courtship procedures lead to two different stable matching systems? If so, what are they?

6. Given the preference structure:

	A	B	C
a	3	1	1
b	2	2	2
c	0	2	3
d	1	3	4

	A	B	C
a	1	2	0
b	1	1	2
c	2	1	3
d	0	2	1

Find all possible stable matching systems, using the Gale–Shapley algorithm.

7. Given the preference structure:

	A	B	C	D	E
a	1	3	5	1	1
b	2	2	4	3	3
c	3	4	3	2	5
d	4	0	2	4	2
e	5	1	1	0	4

	A	B	C	D	E
a	0	2	1	3	0
b	3	2	1	4	5
c	2	1	0	3	4
d	3	1	2	5	4
e	2	3	1	4	5

(1) Find an optimal matching system for the men.

(2) Show that for the above preference structure, this is the optimal matching system for the women too.

(3) Are there other stable matching systems for the above preference structure? Support your answer.

2 Social Justice

2.1 PRESENTATION OF THE PROBLEM

In democratic society, the prevalent method of decision-making is majority rule. This method attempts to aggregate many individual views and opinions into a single social decision.

Suppose there is a community of three voters who must make a decision by choosing one of three alternatives (say, disarmament, cold war, or open war). A society that behaves rationally will establish a preference order with regard to the three alternatives on the basis of voter preferences, choosing the alternative that is the first preference. If, for example, the society establishes a preference order in which the first preference is disarmament, the second preference is cold war, and the third preference is open war, the choice will be for disarmament.

Majority rule is the natural way to make a social decision on the basis of voter preferences.

Consider the following example, known as the "voting paradox."

A certain amount of the municipal budget is unspent and the city council must decide how to invest it. It has three options: investment in education, investment in security, investment in health. (The sum is too small to divide feasibly among the three options.)

Sitting on the city council are representatives of three parties:

Left party: 3 members
Center party: 4 members
Right party: 5 members

The parties' list of preferences is:

Center (4)	**Left (3)**	**Right (5)**
health	education	security
security	health	education
education	security	health

The preferences are listed in the columns in descending order. For example, the Right party prefers to invest the money in security. Its second preference is investment in education, while its third preference is investment in health.

It makes little sense to vote on all the alternatives together in one vote. Such a vote would result in a decision to invest in security (5 to 3 or 4), whereas there is a clear majority in favor of health over security (7 to 5), because both the Left party and the Center party prefer investment in health to investment in security. Therefore, a proposal is adopted to vote on the different alternatives in pairs ("pairwise voting").

"Security" vs. "education" – the majority is in favor of "security" (9 to 3).
"Security" vs. "health" – the majority is in favor of "health" (7 to 5).

In other words, the majority prefers "health" to "security," and it prefers "security" to "education." It therefore seems that the city council will prefer "health" to "security" and "security" to "education."

One may therefore conclude that the social preference is:

health
security
education

However, one of the council members calls for a vote on "health" vs. "education." Remarkably, it turns out that in the majority opinion, "education" is preferred to "health" (8 to 4). In this example, decision by majority leads to the absurd:

health	security	education
security	education	health

The voting results show that health is preferred to security, security is preferred to education, and education is preferred to health. We shall formulate this using a preference symbol:

health $\succ$ security $\succ$ education $\succ$ health

This relation is a cyclic preference relation because, for any alternative, there exists another alternative that is preferred to it. That is, there is no most-preferred alternative and therefore majority rule provides no clear guidance as to how to spend the budget.

This paradox has long been known. The French mathematician and philosopher Marquis de Condorcet first noted it in 1785.

It seems that in this example majority rule, which establishes a social preference on the basis of voter preferences, does not yield rational behavior.

Let us consider another way of making a social decision. The social decision on how to spend the rest of the budget will depend on the relative power of the parties.

security	$\frac{5}{12}$
health	$\frac{4}{12}$
education	$\frac{3}{12}$

Explain why this proposal will be rejected by a majority.

Another option is that the social decision will be whatever the most powerful party dictates.

security
education
health

Does this meet your intuition of a just decision?

The question is whether there is a decision-making model for society that can aggregate known personal preferences in a way that will meet our intuitive demands for a just method of decision-making.

The American economist Kenneth Arrow[1] tried to answer this question. In this chapter we shall discuss the results of his research.

2.2 MATHEMATICAL DESCRIPTION OF THE PROBLEM

In human society, different individuals are likely to have different preference orders with regard to a choice of alternatives. We shall indicate the alternatives by lowercase letters: x, y, z ..., and the individuals by numbers: 1, 2, 3 ... .

The preference orders of the individuals in the society can be arranged in columns, as in the following example:

1	**2**	**3**	**4**
x	y	x~y	t
y	x	z	x~z
z	t	t	y
t	z		

In this example, 1's first preference is alternative x, followed by y, and so on. In contrast, 2's first preference is alternative y, followed by x, and so on. 3 is indifferent to a choice between x and y, but he wants one of them to be his first preference. 4's first preference is t, he is indifferent to a choice between x and z, but he wants one of them to be his second preference, and so on.

This way of listing preferences involves several implicit assumptions, most of them reasonable.

Assumption A:
It is impossible for anyone to prefer x to x.

Assumption B: Transitivity of the preference relation
If someone prefers x to y and y to z, then he prefers x to z.

[1] Arrow, K. J. 1951. *Social choice and individual values.* New York: J. Wiley

From assumption A it follows that x cannot appear at different places in the same column. From assumption B it follows that the listing of preferences in descending order of importance describes the preference order not only between two consecutive alternatives in the column, but also between nonconsecutive alternatives in the column.

Assumption A is very intuitive. We would be puzzled were someone to tell us that he prefers this chocolate cake to this same chocolate cake. Assumption B, however, is less intuitive. For example, why is it that one would not say that he prefers brownies to cake, cake to ice cream, and ice cream to brownies?

In fact, experience shows that if someone is asked many questions about his preferences, his answers often reveal a lack of transitivity. The question is whether his answers reflect his actual preference order or whether that is not the case because he has not thought things through properly.

To summarize: when someone describes preferences that lack transitivity, it is far from clear what he means by the word "prefer." For our part, we shall work from the assumption that everyone has a preference order that confirms the transitivity property.

Assumption C: Asymmetry
If someone prefers x to y, then he does not prefer y to x.

When someone says that he prefers x to y, it is clear that he does not prefer y to x.

This way of listing two or more alternatives in the same row in a column of preferences involves additional implicit assumptions, all tied to the concept of indifference. We shall formulate four of them:

Assumption D: The reflexive relation of indifference
It is impossible for anyone not to be indifferent between x and x.

Assumption E: The transitive relation of indifference
If someone is indifferent to a choice between x and y and he is indifferent to a choice between y and z, then he is indifferent to a choice between x and z.

Assumption F:
If someone prefers x to y and is indifferent to a choice between y and z, then he prefers x to z.

Assumption G:
If someone prefers y to x and is indifferent to a choice between y and z, then he prefers z to x.

The following assumption is the most important.

Assumption H: The Completeness Assumption:
For any two alternatives x and y, there exists exactly one of the following preference orders: the preference for x over y, the preference for y over x, indifference to a choice between x and y.

If this assumption did not exist, then there would be alternatives that we would not know how to include in a preference order.

2.3 EXERCISES

1. At the end of the school year the class committee still has a small amount of money in its petty cash fund. The committee must decide whether to return the money to the students, buy theater tickets for the entire class, or use the money to organize an end-of-the-year party. The amount of money at the committee's disposal is enough to finance only one of these objectives. There are three members on the committee: 1, 2, 3.

The preference orders of the committee members with regard to the alternatives are: Member 1 prefers returning the money to the students to buying theater tickets, but prefers buying theater tickets to organizing an end-of-the-year party. Member 2 prefers organizing an end-of-the-year party to returning the money, but prefers buying theater tickets to organizing an end-of-the-year party. Member 3 prefers organizing an end-of-the-year party to buying theater tickets, but prefers buying theater tickets to returning the money.

(1) Let x, y, and z be the alternatives. List the preference orders of the committee members with regard to these alternatives.

(2) What will be the social decision if it is decided to put the alternatives to a pairwise majority vote?

2. Allen was asked what his fast food preferences were. His answer was: "I prefer a pizza to a sandwich; I prefer a hamburger to a sandwich; I prefer a burrito to a pizza; I prefer a taco to a hamburger; I prefer a falafel to a burrito; I prefer a hamburger to a falafel."

(1) Let p, s, h, b, t, and f (short for pizza, sandwich, hamburger, burrito, taco, and falafel, respectively) be the different lunches mentioned by Allen. Try to list Allen's preferences in a column.
(2) Does the information given by Allen enable us to list his preferences in a column in a way that confirms all the assumptions described in Section 2.2?
(3) Is there any superfluous information? If so, what?

3. Michael described how he spends his leisure hours. He prefers movies to theater; he prefers theater to dancing; he prefers dancing to a show; he prefers reading to concerts; he prefers TV to reading; he prefers TV to theater; he prefers concerts to movies.

(1) Is it possible to list his preferences in a column that will represent his taste in leisure activities?
(2) Is there any superfluous information? If so, what?

4. Sarah told us her music preferences. She prefers Bach to Mozart; she prefers Brahms to Schumann; she is indifferent to a choice between Schumann and Chopin; she is indifferent to a choice between Mozart and Brahms.

What are Sarah's preferences, when she must choose between Bach and Chopin?

5. Jacob tells his son he is making eggs for breakfast and asks him whether he would like his scrambled, hard-boiled, soft-boiled, sunny-side up or sunny-side down. His son replies: "I prefer scrambled to sunny-side up; I prefer sunny-side down to sunny-side up; I prefer

sunny-side up to hard-boiled; I prefer hard-boiled to scrambled; I prefer sunny-side down to soft-boiled."

(1) Does Jacob have enough information to know his son's first preference?
(2) If this information is incomplete, what would you ask in order to complete it?
(3) Does this information contradict any of the assumptions about preference order? If so, indicate the contradiction.

6. Gail told her friends that she loves cinema. Her friends asked her, "If you could take only one movie with you to a desert island, would you take an action film, a horror film, a comedy, a western, or a science-fiction film?"

Gail answered: "I prefer an action film to a horror film; I prefer a science-fiction film to a horror film; I prefer a horror film to a comedy; I prefer a science-fiction film to a western; I prefer a comedy to an action film; I prefer a horror film to a western; I prefer a western to an action film."

(1) Given this information, try to determine what movie Gail would take with her to a desert island.
(2) If this information is incomplete, ask Gail only the questions needed to complete the information.
(3) Does Gail's information contradict any of the assumptions about preference order? If so, what is the contradiction?

7. A host is ready to offer his guest any of the following drinks: coffee, tea, chocolate milk, cappuccino, milk. The guest's preferences are: he is indifferent to a choice between coffee and cappuccino; he is indifferent to a choice between tea and milk; he prefers chocolate milk to tea; he prefers cappuccino to chocolate milk.

(1) What will the host offer?
(2) What are the guest's preferences, when he must choose between coffee and milk?

(3) What are the guest's preferences, when he must choose between cappuccino and tea?

2.4 SOCIAL CHOICE FUNCTION

We have focused on decision-making in society when various alternatives are under discussion. The society can be a parliament, a city council, a board of directors, and so on. Depending on the "society" in question, the components of the society can be political parties, corporations, individuals, and so on.

Every component of society has a complete preference order with regard to the various alternatives under discussion that satisfy Assumptions A–F. An aggregate of preference orders, compiled as described in Section 2.2, is called a *preference profile*. For example:

1	**2**	**3**
x	x~t	y
y	z	x
z	y	t
t		z

Our task is to look for a *decision rule* that assigns to each preference profile a preference order that will represent the social decision. In mathematical terms, we are looking for a *function*, denoted f, that assigns one preference order to each preference profile. Such a function is called a *social choice function*.

Note: The task we set ourselves is not only to choose the most-preferred alternative from among all alternatives under discussion, but also to determine the preference order of the society with regard to all the alternatives under discussion.

There is an advantage in knowing the whole preference order, rather than knowing just the most-preferred alternative. Suppose we decide

x

that the order is y and then it turns out to be impossible to realize x;

z

in such a case it will be possible to choose y, with no need for a revote.

Following are a few examples of social decision rules. First we shall evaluate them against an intuitive standard of "justice." Then we shall present formal conditions for defining the concept of "justice."

I. Majority Rule: f is a decision by a pairwise majority vote.

Example 1

$$f\begin{pmatrix} x & x & t \\ y & t & z \\ z & z & y \\ t & y & x \end{pmatrix} = \begin{pmatrix} x \\ t \\ z \\ y \end{pmatrix}$$

Explanation: A look at the different pairs reveals that society will prefer x to y, because two persons prefer x to y and only one prefers something else. Grouping the preferences in pairs we get:

$$\begin{matrix} x & x & x & z & t & t \\ y & z & t & y & z & y \end{matrix}$$

In the list of pairs we see that x is preferred to all the alternatives and therefore society will rank it as the first preference. t is preferred to all the alternatives except x and therefore it is ranked as the second preference. This leaves z and y; between them, z is preferred to y, and therefore y is ranked as the last preference. Thus we arrive at the social decision above.

Example 2

$$f\begin{pmatrix} x & x{\sim}y & y \\ y & t & x \\ z & z & t \\ t & & z \end{pmatrix} = \begin{pmatrix} x{\sim}y \\ t \\ z \end{pmatrix}$$

In this example society is indifferent to a choice between x and y, because one individual prefers x to y and one individual prefers y to x, while the third is indifferent to a choice between them. In the choice between t and z, two individuals prefer t to z and therefore society as a whole prefers t to z. Everyone prefers x to t and thus we arrive at the social decision above.

Example 3

$$f\begin{pmatrix} x & y & z \\ y & z & x \\ z & x & y \end{pmatrix}$$

In this example, discussed in Section 2.1 (p. 61), f gives no decisive answer concerning the social decision. According to *f*:

$$\begin{matrix} x & y & z \\ y & z & x \end{matrix}$$

Here, it is impossible to list the preferences in a column. In other words, in this example majority rule does not yield a transitive preference order.

Exercise: Given the following preference profiles, what will be the social decision when the guiding rule is to decide by a pairwise majority vote?

(1)

$$f\begin{pmatrix} x & t & y \\ y & x & x \\ z & z & t \\ t & y & z \end{pmatrix} =$$

(2)

$$f\begin{pmatrix} x\sim y & z & t \\ t & x & z\sim x \\ z & t & y \\ & y & \end{pmatrix} =$$

II. Constant Function

$$f\begin{pmatrix}\text{any}\\\text{profile}\end{pmatrix} = \begin{pmatrix}x\\y\\z\\t\end{pmatrix}$$

This function assigns the preference order $\begin{matrix}x\\y\\z\\t\end{matrix}$ to every preference profile without regard for the individual preferences of the community members. What do you think about this rule as a solution for decision-making in a democracy?

III. Decision by a Dictator

According to this decision rule, a social decision is made in accordance with the will of one individual (whose preferences are here listed in the first column). We call him a *dictator*.

$$f\begin{pmatrix}x & t & t\\y & z & x\\z & y & z\\t & x & y\end{pmatrix} = \begin{pmatrix}x\\y\\z\\t\end{pmatrix}$$

↑
dictator

Does this rule meet your intuition of justice in a democracy?

Exercise: Given the following preference profile, what will be the social decision when the guiding rule is to decide in accordance with the will of one individual (whose preferences are here listed in the second column)?

$$f\begin{pmatrix}x & t & x & x\\y & z & t & y\\z & y & y & t\\t & x & z & z\end{pmatrix} =$$

IV. Ill-defined "Rule"

The opinion of the first individual will decide in a choice between pairs of alternatives x, y, and z. As for the pairs that include alternative t, the opinion of the second individual will decide. For example:

Example 1

$$f\begin{pmatrix} x & t & x \\ y & x & t \\ z & z & y \\ t & y & z \end{pmatrix} = \begin{pmatrix} t \\ x \\ y \\ z \end{pmatrix}$$

Example 2

$$f\begin{pmatrix} x & y & x \\ y & x & t \\ z & t & y \\ t & z & z \end{pmatrix} = \begin{pmatrix} x \\ y \\ t \\ z \end{pmatrix}$$

It seems that this rule always leads to a social decision. But it is not so.

What do you think will happen, given the following preference profile?

Example 3

$$f\begin{pmatrix} y & x & x \\ x & t & y \\ t & y & z \\ z & z & t \end{pmatrix} = ?!$$

What is unattractive about this rule?

Exercise: Given the following preference profiles, what will be the social decision when the guiding rule is: the opinion of the first individual will decide in a choice between pairs of alternatives x, y, and z;

as for the pairs that include alternative t, the opinion of the second individual will decide?

(1)

$$f\begin{pmatrix} x & y & z \\ t & x & x \\ z & t & t \\ y & z & y \end{pmatrix} =$$

(2)

$$f\begin{pmatrix} t & t & y \\ z & x & z \\ x & y & t \\ y & z & x \end{pmatrix} =$$

V. "Just" Rule

Here is a rule that seems just where a "society" of two individuals is concerned. If, for example, both prefer x to y, then society prefers x to y. However, if one prefers x to y and the other prefers y to x, the social decision will be indifferent to a choice between x and y. (We leave it to the reader to decide what the decision will be if one of the individuals prefers x to y and the other is indifferent to a choice between them, and what the decision will be if both are indifferent to a choice between x and y.)

Let us see what society will decide in the following example:

$$f\begin{pmatrix} x & z \\ y & x \\ z & y \end{pmatrix} = ?$$

No doubt, society will decide that x is preferred to y, because this is the unanimous opinion.

There is indifference to the choice between x and z and likewise to the choice between y and z, because society is divided in its opinions about these alternatives.

We shall now try to formulate the example above as a social decision:

x

y~z~x

But this is impossible, because (by the transitivity of preference/indifference relations) x is preferred to x, which of course is impossible.
Therefore, the "just" rule does not establish a social decision for every preference profile.

Exercise 1: Given the following preference profiles, what will be the social decision when the society consists of two individuals and the guiding rule is: if two individuals prefer x to y, then society prefers x to y; if they are divided in their opinions (that is, if one individual prefers x to y and the other individual prefers y to x), then society will be indifferent to a choice between x and y?

(1)

$$f\begin{pmatrix} x & z \\ z & x \\ y & y \end{pmatrix} =$$

(2)

$$f\begin{pmatrix} z & x \\ y & y \\ x & z \end{pmatrix} =$$

(3)

$$f\begin{pmatrix} y & y \\ x & z \\ z & x \end{pmatrix} =$$

(4)

$$f\begin{pmatrix} z & y \\ x & z \\ y & x \end{pmatrix} =$$

Exercise 2: Given the following preference profiles, what will be the social decision when the society consists of two individuals and the guiding rule is: if two individuals prefer x to y, then society prefers x to y; if both are indifferent to a choice between x and y, then society is indifferent to a choice between x and y; if they are divided in their opinions, then society is indifferent to a choice between x and y; and if one individual prefers x to y and the other individual is indifferent to a choice between them, then society prefers x to y?

(1)

$$f\begin{pmatrix} x\sim y & x \\ z & z\sim y \end{pmatrix} =$$

(2)

$$f\begin{pmatrix} z & x \\ x\sim y & y \\ & z \end{pmatrix} =$$

VI. Dependence on Irrelevant Alternatives

What is wrong with the following rule involving a society of three individuals and three alternatives?

Rule: A decision must be made by majority rule; if the result is a cyclic preference relation (p. 61), society must decide on the alternatives in alphabetical order.

Consider the following two examples:

Example 1

$$f\begin{pmatrix} x & y & z \\ y & z & x \\ z & x & y \end{pmatrix} = \begin{pmatrix} x \\ y \\ z \end{pmatrix}$$

Explanation: Majority rule establishes a cyclic order $\begin{matrix} x \\ y \\ z \\ x \end{matrix}$ and therefore the social decision is made according to alphabetical order.

Example 2

$$f\begin{pmatrix} y & y & y \\ x & z & z \\ z & x & x \end{pmatrix} = \begin{pmatrix} y \\ z \\ x \end{pmatrix}$$

The interesting thing about this rule is that insofar as alternatives x and z are concerned, there is no difference between the preference relations in the two examples (the first individual prefers x to z and the other two prefer z to x). Nevertheless, the social decision between x and z is different in the two cases. In the second example the preference order changes in favor of y. Indeed, in this example society makes y its first preference. But the change in the place of y in the preference order affects the social preference with regard to the choice between x and z.

Is independence from irrelevant alternatives necessary or not?

Exercise: (1) Given the following preference profiles, what will be the social decision when the guiding rule is to decide by a pairwise majority vote; if the result is a cyclic preference relation, then society will decide according to reverse alphabetical order?

(i)

$$f\begin{pmatrix} y & z & x \\ z & x & y \\ x & y & z \end{pmatrix} =$$

(ii)

$$f\begin{pmatrix} z & z & z \\ y & x & x \\ x & y & y \end{pmatrix} =$$

(2) What is the difference between preference profile (i) and preference profile (ii)?

(3) How does this difference affect the social decision?

(4) Why is this decision rule not recommended?

VII. Positive Association of Individual Preferences and Social Preferences

What is wrong with the following rule involving a society in which alternatives x and y are voted on pairwise?

Rule: If the majority that prefers x to y consists of an even number of individuals, then society prefers x to y. If the majority that prefers x to y consists of an odd number of individuals, then society prefers y to x.

Example 1

$$f\begin{pmatrix} x & x & x & x & y \\ y & y & y & y & x \end{pmatrix} = \begin{pmatrix} x \\ y \end{pmatrix}$$

However,

Example 2

$$f\begin{pmatrix} x & x & x & x & x \\ y & y & y & y & y \end{pmatrix} = \begin{pmatrix} y \\ x \end{pmatrix}$$

In the first example, four out of five individuals prefer x to y and therefore, according to the rule, society prefers x to y.

In the second example, five out of five individuals prefer x to y and therefore, according to the rule, society prefers y to x.

Does this rule meet your intuition of justice in a democracy?

Exercise: (1) Given the following preference profiles, what will be the social decision when the guiding rule is: if the majority that prefers x to y consists of an even number of individuals, then society prefers x to y; if the majority that prefers x to y consists of an odd number of individuals, then society prefers y to x?

(i)

$$f\begin{pmatrix} x & z & y & z & x \\ y & x & x & y & z \\ z & y & z & x & y \end{pmatrix} =$$

(ii)

$$f\begin{pmatrix} x & z & x & z & x \\ y & x & y & x & z \\ z & y & z & y & y \end{pmatrix} =$$

(2) What is the difference between preference profile (i) and preference profile (ii)?

(3) Is there a difference in the social preference in the two examples?

(4) Why is this decision rule not recommended?

2.5 AXIOMS FOR THE SOCIAL CHOICE FUNCTION

In this section we shall try to find a social choice function without the disadvantages discussed in Section 2.4.

First we shall formulate requirements for f; then we shall look for a function f that satisfies these requirements. These requirements are called axioms, and they are the formal conditions for defining the properties of a just decision rule.

Axiom 1: The Domain and Range of the Function

The domain of the function f is all possible preference profiles of the components of society and its range is the preference orders with regard to the alternatives under discussion; henceforth the image of f will be called the "social preference."

$$f\begin{pmatrix} \text{preference} \\ \text{profile} \end{pmatrix} = \begin{pmatrix} \text{social} \\ \text{preference} \end{pmatrix}$$

This axiom prevents us from implementing the following rules discussed in Section 2.4: Rule I (the majority rule, p. 69, Example 3), Rule IV (the "ill-defined" rule, p. 71, Example 3), and Rule V (the "just" rule, p. 72).

In all these examples there were preference profiles for which no social preference was obtained.

Axiom 2: Positive Association of Individual Preferences and the Social Preference
Let P and Q be two preference profiles in which preference or indifference relations are the same for all pairs of alternatives except x and y, but preference or indifference relations that concern x and y "favor" x in Q. In this case, the social choice function f that concerns x and y is either as good or better for x in Q.

Explanation: "Preference or indifference relations that concern x and y favor x in Q" means that if an individual prefers x to y in P, then he also prefers x to y in Q. If he is indifferent between x and y in P, then either he is indifferent between x and y or he prefers x to y in Q. If he prefers y to x in P, then either he prefers y to x, is indifferent between y and x, or prefers x to y in Q. Similarly, "the social choice function *f* that concerns x and y is either as good or better for x in Q" means that if the social choice function *f* of preference profile P determines that x is preferred to y, then the social choice function of profile Q also determines that x is preferred to y. If the social choice function *f* of preference profile P determines that x is indifferent to y, then the social choice function of profile Q determines that either x is indifferent to y or x is preferred to y.

Suppose members of a parliament declare their preference orders with regard to all alternatives discussed on a given day, and it is decided that alternative x is preferred to alternative y. On the following day the parliament members are asked again to declare their preference orders with regard to the exact same alternatives. In their new preference profile there is a change that affects only the pair of alternatives x and y: some individuals now rank x higher in their preference order than they did on the day before. It would be strange if after this change parliament decided that y was preferred to x. Axiom 2 aims at preventing such a possibility.

According to this axiom, if given a certain preference profile the social decision is that x is preferred to y, then that will be the decision all the more given a preference profile where none of the individuals rank x lower in their preference order relative to the original preference profile, assuming everything remains unchanged in every pair of alternatives other than x and y.

Rule VII in Section 2.4 (p. 76) does not satisfy this axiom.

Axiom 3: Unanimous Decision
If all individuals in a society prefer x to y, then the social decision will prefer x to y.

Explanation: It would be strange if everybody preferred x to y and f established that y is preferred to x, or that there is indifference to a choice between x and y. Axiom 3 aims at preventing such a possibility.

Rule II in Section 2.4 (p. 70) does not satisfy this axiom.

Axiom 4: Independence of Irrelevant Alternatives
The social choice function f concerning preference/indifference relations between x and y depends only on what members of society think about the preference relation between x and y. What they think about the association between z and t or even between x and t is irrelevant to the association between x and y.

Explanation: Let us imagine a committee asked to elect a chairman. There are two candidates for the position: candidate A and candidate B. It emerges from a discussion in which each committee member declares his preference that A will be elected chairman. But someone in the committee suddenly remembers that C is a candidate too. It would be strange if the group then decided that B will be the chairman. The presence or absence of candidate C is irrelevant to the decision between candidates A and B.

Axiom 4 requires that the social choice function be such that irrelevant alternatives will not affect the preferences under discussion.

Rule VI in Section 2.4 (p. 74) does not satisfy this axiom.

Axiom 5: Non-dictatorship
In a society of at least three individuals there is no dictator; that is, there is no individual whose opinion decides all issues even if everyone else opposes his opinion.

Explanation: This axiom is concerned with an individual whose opinion decides *all issues*. We prohibit such a possibility, but we certainly permit an individual's opinion to decide some issues. In that case he is not a dictator.

Rule III in Section 2.4 (p. 70) does not satisfy this axiom.

2.6 EXERCISES

1. A social choice function for a given preference profile is:

$$f\begin{pmatrix} x & z & x \\ y & x & y \\ z & y & z \end{pmatrix} = \begin{pmatrix} y \\ x \\ z \end{pmatrix}$$

Which axiom is not satisfied here?

2. (1) Find a social choice function for the following preference profile according to a pairwise majority vote:

$$f\begin{pmatrix} x & y & z \\ y & z & x \\ z & x & y \end{pmatrix} =$$

(2) Which axiom is not satisfied here?

3. (1) Find a social choice function for the following preference profiles according to the following rule: if an odd number of people prefer x to y, society will prefer x to y; if an even number of people prefer x to y, society will prefer y to x. As for the other alternatives, they are decided by a pairwise majority vote and in case of a tie there is indifference.

$$f\begin{pmatrix} x & t & z & y \\ y & x & x & x \\ z & y & y & t \\ t & z & t & z \end{pmatrix} =$$

$$f\begin{pmatrix} x & t & z & x \\ y & x & x & y \\ z & y & y & t \\ t & z & t & z \end{pmatrix} =$$

(2) Which axiom is not satisfied here?

4. (1) Find a social choice function for the following preference profiles according to the following rule: we are trying to decide by a pairwise majority vote; if the outcome is a cyclic preference relation, society decides according to reverse alphabetical order.

$$f\begin{pmatrix} z & y & x \\ y & x & z \\ x & z & y \end{pmatrix} =$$

$$f\begin{pmatrix} y & y & y \\ z & x & x \\ x & z & z \end{pmatrix} =$$

(2) What is the difference between the two preference profiles above?

(3) Which axiom is not satisfied here?

5. A social choice function for a given preference profile is:

$$f\begin{pmatrix} x & t & z & y \\ y & y & x & t \\ z & z & t & x \\ t & x & y & z \end{pmatrix} = \begin{pmatrix} z \\ x \\ t \\ y \end{pmatrix}$$

Which axiom is not satisfied here?

2.7 WHAT FOLLOWS FROM AXIOMS 1–4?

In the previous sections we introduced several requirements for the social choice function f. They are all simple and reasonable. We called these requirements *axioms*. In this section we shall see what conclusions can be drawn only from Axioms 1–4 concerning the function f.

The Axioms:

1. The function f assigns a preference order, called a *social preference*, to every preference profile.
2. The function f reflects a positive association between the preferences of the individuals in society and the social preference: if in a certain preference profile f establishes the preference order $\begin{smallmatrix} x \\ y \end{smallmatrix}$ and the preference profile changes so that more individuals in society prefer x to y, or $\begin{smallmatrix} y \\ x \end{smallmatrix}$ becomes $y \sim x$, then f establishes that x is preferred to y in the new preference profile too.
3. The function f obeys a unanimous choice: if all the individuals in society prefer x to y, f also establishes that x is preferred to y.
4. The function f establishes a preference between alternatives x and y independently of any other alternatives; that is, what f establishes depends on the preference orders of the individuals in society with regard only to x and y.

Examples:

1. Given that f satisfies the above axioms, what will f establish as the social preference if the preference profile of a society of three individuals is:

1	**2**	**3**
x	z	x
y	x	z
z	y	y
t	t	t

Solution:

The function f will establish the preference order $\begin{smallmatrix} x \\ y \end{smallmatrix}$ because everybody prefers it (unanimous decision axiom). For the same reason, f will establish the preference relations $\begin{smallmatrix} z \\ t \end{smallmatrix}\ \begin{smallmatrix} y \\ t \end{smallmatrix}\ \begin{smallmatrix} x \\ t \end{smallmatrix}$; hence, by the unanimous decision axiom, f will establish the preference orders $\begin{smallmatrix} x \\ y \end{smallmatrix}\ \begin{smallmatrix} z \\ t \end{smallmatrix}\ \begin{smallmatrix} y \\ t \end{smallmatrix}\ \begin{smallmatrix} x \\ t \end{smallmatrix}$.

The independence of irrelevant alternatives axiom applies here as well, because the preferences described above were established independently of the preferences of the individuals in society with regard to the other alternatives.

For example, $\begin{matrix} x \\ y \end{matrix}$ is independent of the preference orders with regard to z and t; $\begin{matrix} z \\ t \end{matrix}$ is independent of the preference orders with regard to x and y, and so on.

We still do not know what *f* will establish either with regard to alternatives x and z or with regard to alternatives y and z, because as far as these alternatives are concerned society is divided in its opinions.

If we knew more properties of *f*, we could perhaps say more. For example, if we assumed that *f* establishes the preference order $\begin{matrix} y \\ z \end{matrix}$, we could conclude, by the transitivity property of the preference relation, that *f* also establishes the preference order $\begin{matrix} x \\ z \end{matrix}$. That is, we would have that $\begin{matrix} x & y & z & y & x \\ y & z & t & t & t \end{matrix}$, and therefore we could say that:

$$f\begin{pmatrix} x & z & x \\ y & x & z \\ z & y & y \\ t & t & t \end{pmatrix} = \begin{pmatrix} x \\ y \\ z \\ t \end{pmatrix}$$

and thus establish *f* for this example.

2. Given that *f* satisfies the above axioms:

   I. What will *f* establish if the preference profile in a society of three individuals is:

1	**2**	**3**
x	z	x
z	x	y
y	y	z

Solution:
By the unanimous decision axiom, we can say that f will establish as the social preference that x is preferred to y, because all individuals in society prefer it. Moreover, the independence of irrelevant alternatives axiom indirectly applies here, because this preference was established independently of society's preferences with regard to z. The given preference profile does not enable us to say anything about the social preference with regard to the pairs x, z or y, z, because society is divided in its opinions with regard to these alternatives: some individuals prefer one possibility and others prefer the other.

II. Given the same preference profile, what can you say about f if it is known that f establishes that y is preferred to z?

Solution:
We know that: $\begin{matrix} x & y \\ y & z \end{matrix}$. By the transitivity of the preference relation, f must also establish that x is preferred to z.

$$f\begin{pmatrix} x & z & x \\ z & x & y \\ y & y & z \end{pmatrix} = \begin{pmatrix} x \\ y \\ z \end{pmatrix}$$

This example is of interest because it illustrates a *prediction*. Although we did not know in advance what f would decide with regard to alternatives x and z, we have managed to predict the social decision.

Remark:
If additional information were to reveal that, say, f establishes a preference for z over y, we would not be able to predict the social decision. In this case the information would amount to $\begin{matrix} x & z \\ y & y \end{matrix}$, from which it is impossible to conclude how f would order the pair of alternatives x, z.

3. Given that f satisfies all four axioms, describe what the social decision will be given the following preference profile, if it is also

known that f establishes a preference for y over z.

1	**2**	**3**
x	z	x
y	x	z
z	y	y
t	t	t

Solution:

In many cases it is possible to know what f establishes, like when all individuals in society prefer a certain alternative to another alternative (unanimous decision). Let us compile all the available information:

x	x	y	z	y
y	t	t	t	z

The first four columns of preferences from the left were obtained by the unanimous decision axiom. The last column is additional information. According to the available information and by the transitivity of the preference relation, we get $\begin{matrix} x \\ z \end{matrix}$. Now we have all the information and we may conclude that:

$$f\begin{pmatrix} x & z & x \\ y & x & z \\ z & y & y \\ t & t & t \end{pmatrix} = \begin{pmatrix} x \\ y \\ z \\ t \end{pmatrix}$$

2.8 EXERCISES

1. (1) Describe the social decision given the following preference profile:

1	**2**
x	z
t	x
y	t
z	y

(2) If we know that f establishes a preference for x over z, what will the social decision be?

(3) What other information must we have in order to know what the social decision will be?

2. (1) Can we predict the social decision given the following preference profile?

1	2	3
x	z	y
y	x	x
z	y	z

(2) If we know that the social decision establishes a preference for x over y, what will the social choice function be?

(3) Can we predict the social decision given the following preference profile if, besides the information in (2), we also know that f establishes $\begin{matrix} y \\ z \end{matrix}$?

1	2	3
x	x	y
y	z	x
z	y	z

3. (1) Can we predict the social decision given the following preference profile?

1	2	3
x	x	x
z	z	z
y	y	t
t	t	y

If so, state the social decision; if not, provide the missing information.

(2) Describe the social decision given the preference profile above, when it is known that f establishes a preference for t over y.

(3) Describe the social decision given the preference profile above, when it is known that f establishes a preference for y over t.

2.9 ARROW'S THEOREM

The present chapter opened with a discussion of the shortcomings of majority rule as a principal method of social decision-making. The question that was raised in light of these shortcomings was whether it is possible to find another, "fair" way of making a social decision. In the course of this chapter we constructed a system of axioms, that is, a system of intuitive requirements for a fair decision-making procedure. The question now is whether there is a social decision rule for all possible preference profiles that satisfies this system of axioms.

Kenneth Arrow's surprising answer is that there is no social choice function that satisfies all the axioms! This means that every social choice function we can think of will fail to satisfy at least one of the axioms. In other words, there is an internal contradiction in the system of axioms presented in this chapter.

In this section we shall prove Arrow's theorem about the non-existence of a social choice function that satisfies all the axioms. In the course of the proof we shall see that any decision rule that does satisfy Axioms 1–4 must be a dictatorial decision rule, which contradicts Axiom 5, defined in Section 2.5 (p. 79).

For the proof, we assume that a decision rule satisfies Axioms 1–4.

Definition:

A set of individuals V is said to be *decisive for the pair* $\genfrac{}{}{0pt}{}{x}{y}$ if, *for every preference profile* in which everyone in V prefers $\genfrac{}{}{0pt}{}{x}{y}$ and everyone else prefers $\genfrac{}{}{0pt}{}{y}{x}$, the social choice function f establishes $\genfrac{}{}{0pt}{}{x}{y}$.

In other words, V is decisive for $\genfrac{}{}{0pt}{}{x}{y}$ if f establishes $\genfrac{}{}{0pt}{}{x}{y}$ whenever the members of V have this preference and the other members have the opposite preference.

Note: If there is a dictator in the society, then he constitutes a decisive set for every pair of alternatives (and not for just one pair).

Remark: According to the definition, V is decisive for $\genfrac{}{}{0pt}{}{x}{y}$ if all members of V prefer $\genfrac{}{}{0pt}{}{x}{y}$ and all the other members prefer $\genfrac{}{}{0pt}{}{y}{x}$. But if some members not in V prefer $\genfrac{}{}{0pt}{}{x}{y}$ or $x \sim y$, then this profile favors $\genfrac{}{}{0pt}{}{x}{y}$ and by Axiom 2, the social choice function will continue to favor $\genfrac{}{}{0pt}{}{x}{y}$. In any case, the preference profile tends to favor x; therefore, by Axiom 2, the social choice function will decide in favor of $\genfrac{}{}{0pt}{}{x}{y}$.

Is There a Decisive Set?

The answer is yes! The set of *all individuals* is no doubt a decisive set, and not only for a certain pair $\genfrac{}{}{0pt}{}{x}{y}$, but for every pair, by the unanimous decision axiom (Axiom 3).

Consider the set of all individuals that, as we said, is decisive for every pair of alternatives. It might be possible for us to subtract some individuals from the set, so that the remaining set will still be decisive – if not for all pairs, then at least for one pair. We shall keep subtracting individuals from the set, as long as there remains a set of individuals that is decisive for some pair. We shall continue the procedure until we have a set that is still decisive for some pair $\genfrac{}{}{0pt}{}{x}{y}$, but from which no more individuals can be subtracted, because that would result in a set that is not decisive for any pair of alternatives. The smallest decisive set for any pair of alternatives is called a *minimal decisive* set.[2]

[2] This set cannot be the empty set, because if the empty set were decisive for the pair $\genfrac{}{}{0pt}{}{x}{y}$, the social choice function would establish $\genfrac{}{}{0pt}{}{x}{y}$ even if everyone preferred $\genfrac{}{}{0pt}{}{y}{x}$, which contradicts the unanimous decision axiom.

Definition:

Set V is called a *minimal decisive set* if V is decisive for a certain pair $\genfrac{}{}{0pt}{}{x}{y}$ and if subtracting individuals from the set results in a set that is not decisive for any pair of alternatives.

We have thus proved that there exists a minimal decisive set; that is, there exists a set that is decisive for a certain pair $\genfrac{}{}{0pt}{}{x}{y}$ and any strict subset of it is not decisive for any pair of alternatives.

Let V be a minimal decisive set. We denote the pair for which it is decisive by $\genfrac{}{}{0pt}{}{x}{y}$. Let j be a specific individual in V and let W be the set consisting of the remaining individuals in V. We denote the set of all individuals not in V by U. Now, consider the following preference profile:

V		
{j}	W	U
x	z	y
y	x	z
z	y	x

For this preference profile the social choice function *f* will establish the preference relation $\genfrac{}{}{0pt}{}{x}{y}$, because all individuals in V = {j}∪W prefer x to y and V is a decisive set for the pair $\genfrac{}{}{0pt}{}{x}{y}$.

The social choice function will not be able to establish $\genfrac{}{}{0pt}{}{z}{y}$, because only the individuals in W prefer it and W is not a decisive set, because V is the minimal decisive set. We apply here the independence of irrelevant alternatives axiom (Axiom 4), which allows us to rule out the possibility that the position of x in the preference relations may affect the preference between y and z.

Hence, the social choice function will establish $\genfrac{}{}{0pt}{}{y}{z}$ or y~z.

Since we already have $\genfrac{}{}{0pt}{}{x}{y}$, *f* will establish the preference relation $\genfrac{}{}{0pt}{}{x}{z}$ by the transitivity of the preference relation. But only j prefers it, while everyone else prefers the contrary. Hence it follows by Axiom 4 that {j} constitutes a decisive set for the pair $\genfrac{}{}{0pt}{}{x}{z}$. But V is a minimal decisive set and therefore $W = \emptyset$ and V={j}.

Moreover, because z denotes any alternative, it follows that {j} is decisive for every pair of alternatives $\genfrac{}{}{0pt}{}{x}{z}$.

We have thus proved that if *f* satisfies Axioms 1–4, then there exists a minimal decisive set consisting of a single individual and it is decisive for the pairs $\genfrac{}{}{0pt}{}{x}{z}$ for some x and any z. If no one remains $W = \emptyset$.

Our task now is to show that a decisive set of one individual is a dictator; that is, if V is a decisive set consisting of a single player, then this player must be decisive for every pair of alternatives and not only for pairs of the $\genfrac{}{}{0pt}{}{x}{z}$ kind. As we said, {j} constitutes a decisive set for every pair of the $\genfrac{}{}{0pt}{}{x}{z}$ kind.

Consider the following preference profile:

{j}	U
w	z
x	w
z	x

f will establish $\genfrac{}{}{0pt}{}{x}{z}$, because {j} is decisive for these pairs.

f will also establish $\genfrac{}{}{0pt}{}{w}{x}$, because everyone prefers it.

By the transitivity of the preference relation, f will establish $\genfrac{}{}{0pt}{}{w}{z}$; that is, {j} is decisive for every pair of alternatives $\genfrac{}{}{0pt}{}{w}{z}$, when $w \neq x$, $z \neq x$.

Finally, consider the following preference profile:

{j}	U
w	z
z	x
x	w

f will establish $\genfrac{}{}{0pt}{}{w}{z}$ when $w \neq x$, $z \neq x$, because {j} is decisive for $\genfrac{}{}{0pt}{}{w}{z}$. f will establish $\genfrac{}{}{0pt}{}{z}{x}$, because everyone prefers it.

By the transitivity of the preference relation, f will establish $\genfrac{}{}{0pt}{}{w}{x}$. But only j prefers it, and so {j} is a decisive set for $\genfrac{}{}{0pt}{}{w}{x}$.

Thus we have proved that:

{j} is decisive for $\genfrac{}{}{0pt}{}{x}{z}$ for all z,

{j} is decisive for $\genfrac{}{}{0pt}{}{w}{z}$ for all w and all z different from x, and

{j} is decisive for $\genfrac{}{}{0pt}{}{w}{x}$ for all w.

Those are all the possibilities. Indeed, {j} is decisive for $\genfrac{}{}{0pt}{}{x}{z}$ for all z, $z \neq x$, because z can be replaced by any alternative. Because {j} is decisive for $\genfrac{}{}{0pt}{}{w}{z}$ for all w and all z different from x, w and z can be replaced by any alternative except x. As for the possibility $w = x$, we know that {j} is decisive for $\genfrac{}{}{0pt}{}{x}{z}$. Also, {j} is decisive for $\genfrac{}{}{0pt}{}{w}{x}$ for all w because w can be replaced by any alternative, and with that we have addressed all the possibilities.

We can therefore sum up by saying that {j} is decisive for every pair of alternatives and therefore j is a dictator!

To summarize, we started from the social choice function that satisfies Axioms 1–4 and proved that it is necessarily a dictatorial rule, which does not satisfy Axiom 5.

That is, there is no social choice function that satisfies the system of Axioms 1–5 in its entirety.

2.10 WHAT NEXT?

Our aim throughout this chapter has been to find a social decision rule that will satisfy our sense of fairness in a democratic society. This aim was not achieved; what is more, it was proved that such a rule does not exist!

What is to be done? What rule are we to follow in making decisions? How is society to conduct its affairs? From all that has been said in this chapter it follows that there are no satisfactory answers to these questions. We must accept the fact that every decision rule that is chosen will not satisfy at least one of Arrow's axioms.

Arrow's book, in which he proved the theorem that now bears his name, stirred debate among social scientists over the implications of the impossibility of a satisfactory decision rule. Social science philosophers suddenly realized that the question "What is good for society?" is not always possible to answer. Arrow's conclusion brought about a radical change in many scientists' perception of the human world around us.

Arrow's book also led to mathematical research. For example, mathematicians wondered whether they could avoid contradicting the axioms by restricting the domain of preference profiles. In fact, narrower domains were established in which all five axioms could be satisfied by appropriate social choice functions. There were also proposals to integrate lotteries into decision rules: if, for example, it were seen that majority rule leads to a cyclic preference relation, then it would be decided by casting lots between the preferences. We shall not go into all that has been done in this area. We shall only note that Arrow's study spawned a huge literature, both theoretical and applied, on this fascinating subject.

2.11 REVIEW EXERCISES

1. Given the following preference profiles, what will the social choice function be when the guiding rule is to decide by a pairwise majority vote?

(1)

$$f\begin{pmatrix} x & t & y \\ y & x & t \\ z & z & x \\ t & y & z \end{pmatrix} =$$

(2)

$$f\begin{pmatrix} x\sim t & t & x\sim z \\ z & x & t \\ y & y & y \\ & z & \end{pmatrix} =$$

2. Given the following preference profiles, what will the social choice function be when society consists of two individuals and the guiding rule is: if both individuals prefer x to y, then society will prefer x to y; if they are divided in their opinions, then society will be indifferent to a choice between x and y; if both individuals are indifferent to a choice between x and y, then society will be indifferent to a choice between x and y; if one prefers x to y and the other is indifferent to a choice between them, then society will prefer x to y?

(1)

$$f\begin{pmatrix} x & y \\ y & z \\ z & x \end{pmatrix} =$$

(2)

$$f\begin{pmatrix} x & z \\ z & x \\ y & y \end{pmatrix} =$$

(3)

$$f\begin{pmatrix} x \sim y & y \\ z & x \sim z \end{pmatrix} =$$

(4)

$$f\begin{pmatrix} y & x \\ z & y \sim z \\ x & \end{pmatrix} =$$

3. (1) Given the following preference profiles, find a social choice function by the following rule: if an even number of individuals prefer x to y, then society will prefer x to y; if an odd number of individuals prefer x to y, then society will prefer y to x. As for the remaining alternatives, they are decided by a pairwise majority vote and in the event of a tie there is indifference.

(i)

$$f\begin{pmatrix} x & z & y & t \\ y & t & x & x \\ t & x & t & y \\ z & y & z & z \end{pmatrix} =$$

(ii)

$$f\begin{pmatrix} x & z & t & y \\ y & x & y & x \\ z & y & x & t \\ t & t & z & z \end{pmatrix} =$$

(iii)

$$f\begin{pmatrix} x & z & t & y \\ y & x & x & x \\ z & y & y & t \\ t & t & z & z \end{pmatrix} =$$

(2) In each case, check whether any axiom is not satisfied. If so, which?

4. Given the following preference profile, describe the social choice function, when it is known that *f* establishes a preference for x over t, and the remaining alternatives are decided by majority rule.

1	2	3
x	t	y
y	x	t
z	z	x
t	y	z

5. (1) Given the following preference profile, describe the social choice function when the guiding rule is to decide by a pairwise majority vote.

1	2	3
x	y	y
y	x	x
z	z	z

(2) What information needs to be added in order to predict the social decision?

6. (1) Given the following preference profile, can the social decision be determined when the guiding rule is to decide by a pairwise majority vote?

1	2	3
x	t	z
y	x	t
z	y	x
t	z	y

(2) If it is known that the social decision establishes a preference for y over z, is this information sufficient to predict the social decision?

(3) Given the following preference profile and the information in (2), can the social decision be determined?

1	**2**	**3**
x	z	z
y	x	t
z	y	x
t	t	y

3 The Shapley Value in Cooperative Games

3.1 INTRODUCTION

Game theory has aroused interest both for its mathematical character and for its many applications to the social sciences. Game theory arises from social phenomena as opposed to physical phenomena. People act, sometimes against each other, sometimes for each other; their interests lead them to conflict or cooperation. By contrast, atoms, molecules, and stars crystallize, collide, and explode, but they do not fight or cooperate. Thus, a mathematical theory was created whose system of concepts is drawn from the social sciences.

The word "game" has different meanings for the layman and the game theorist, but the different meanings have a common denominator: the game has players and the players must interact or make decisions. As a result of the players' actions, and perhaps also by chance, the game will yield a certain outcome that is either a punishment or a reward for each one of the players. The word "player" has an unconventional meaning in that it does not necessarily signify an individual. A player can be a team, a corporation, or a state. It is convenient to refer to a group of persons having a common identifying interest and capable of making joint decisions as a single player. One might say that a *game* is a situation involving several decision-making bodies. Each decision-making body is a *player*.

Human interactions involve many aspects, such as the capabilities of the players, their desires, their values, the role of the environment in which they function, and so on. Game theory selects a few of these aspects and constructs *mathematical models*, usually quite abstract. These models are analyzed and game theory then attempts to provide *recommendations* for behavior and possible resolutions of conflicts. As there are various issues that may be addressed, there may be various types of recommendations. Each such type is called a *solution concept*.

In this chapter we shall study one class of games, called *cooperative transferable utility games*. These are games that involve a division of money among the players, and the rules of the game allow for making *binding agreements*, namely, agreements that will be honored. We shall consider a single solution concept, called the *Shapley value*, which can be regarded as a division of money that a judge or an arbitrator is likely to recommend. This Shapley value also has other interpretations which will be discussed subsequently.

3.2 COOPERATIVE GAMES

Cooperative games are games in which players enter into mutually binding agreements. For example, economic negotiations often conclude in a contract binding on all parties, and the parties are unlikely to break the contract owing to the penalties attaching to such a breach.

In contrast, *non-cooperative games* are games in which players enter into nonbinding agreements. For example, political agreements, like those between states, are generally nonbinding, and the parties to a political agreement honor it only for as long as they see fit.

A large part of the theory of cooperative games deals with *coalition function games*,[1] whose essential features will be discussed in this section.

The mathematical model of a cooperative game is the pair $(N; v)$, where $N = \{1, 2, 3, ..., n\}$ is the set of *players*. The coalition function v will be explained shortly.

Every subset of N is called a *coalition* and is denoted by a capital letter, S, for example. The expression "coalition S was formed" appears often in the description of games. In theory, the meaning of this expression is that all coalition members gave their consent to

[1] More precisely, *coalition function form games with side payments*, because money is often distributed among the players in these games.

the formation of the coalition. In practice, this expression has various meanings. We provide some examples:

1. I went to the store and asked for a loaf of bread (and the grocer agreed to give it to me). We may say that from the moment this (binding) agreement was made a coalition was formed between the grocer and myself.
2. A group of political parties holding a majority after elections decided to form a governing coalition until the next elections. The formation of a coalition here consists of the agreement to share the burden of power.
3. A group of investors decided to found a factory. The agreement of these investors to found a factory means that a coalition was formed between them.

Of course, coalitions usually do not take place in a vacuum. Coalition building usually requires prolonged contact between the parties, intensive negotiations, and a decision-making procedure suited to the type of agreement (e.g., profit sharing). The analysis of how players decide to behave *after* the coalition is formed is the analysis of *solution concepts*. We shall study these in the sequel.

In this chapter we assume that whenever a coalition S is formed, an amount of money $v(S)$ is generated.[2] Thus, v is a function called the "coalition function," and it assigns a real number to every coalition. The number $v(S)$ is called the *coalition worth* of S.

Example:
An advertising agent approaches three individuals, 1, 2, and 3, and asks them to sign an advertisement saying that they use "Sparkle" toothpaste. The agent says that he is interested in obtaining at least two signatures. If 1 and 2 sign, the agent will pay them a total of \$100. If 1 and 3 sign, the agent will pay them a total of \$100. On the other hand, if 2 and 3 sign, the agent will only pay them a total of \$50. If all three agree

[2] We assume that $v(S)$ is independent of the actions taken by the players not in S.

to sign, the agent will pay them a total of \$120. In this example, the formation of a coalition means the agreement of its members to sign an advertisement.

The mathematical model is:[3]

$$N = \{1,2,3\} \qquad \begin{aligned} &v(1,2) = 100\\ &v(1,3) = 100\\ &v(2,3) = 50\\ &v(1,2,3) = 120\\ &v(1) = v(2) = v(3) = 0\\ &v(\emptyset) = 0 \end{aligned}$$

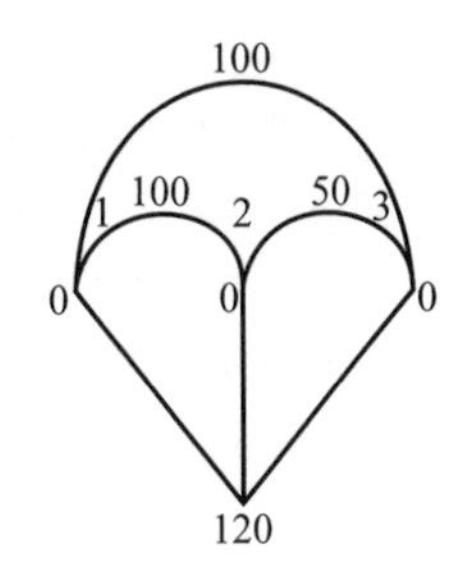

Remarks:

1. A set with one player is called a coalition too (a one-person coalition), because a coalition is a subset of N and sets with one member are also subsets of N.

2. In contrast to everyday language, in game theory it is convenient to regard the empty set as a coalition and to assume that $v(\emptyset) = 0$.

In discussing coalition games, we can focus on how players bargain with each other over how to divide the payoff that they can achieve in a game by banding together and joining forces in a coalition.

We shall now present a possible example of negotiations between players in the game described above:

Player 2 proposes an equal division to player 1:[4] $(50, \mathbf{50}, 0)$
Player 3, who might get nothing, proposes to 1: $(60, 0, \mathbf{40})$

[3] For simplicity, we shall omit curly brackets and write $v(1,2) = 100$, for example, instead of the more precise $v(\{1,2\}) = 100$.

[4] The figure in bold face indicates who made the proposal.

Player 2 lowers his demands so he won't be left out:	$(70, \mathbf{30}, 0)$
Player 3 is prepared to settle for:	$(80, 0, \mathbf{20})$
Player 2 turns to player 3 and proposes:	$(0, \mathbf{25}, 25)$
Player 3 is forced to compete and proposes:	$(70, 0, \mathbf{30})$
Player 2 suggests forming a coalition of all participants with a payoff division of:	$(70, \mathbf{20}, 30)$

If all players reach agreement at this stage, the game terminates, and it is said that the *outcome* of the game is $(70, 20, 30)$ and that the coalition $\{1, 2, 3\}$ was formed.

There is a rich theory that attempts to describe what outcomes are likely to form, but it is beyond the scope of this book. Instead, we address the question of what a judge or an arbitrator is likely to decide if the three players bring the game before him and ask him to propose a "fair" division.

3.3 IMPORTANT EXAMPLES OF COALITION FUNCTION GAMES

Example 1: Two-Person Bargaining Game

$N = \{1, 2\}$ $\qquad v(N) = 1$

$v(1) = v(2) = 0$

In this game the players achieve nothing separately; that is, each player achieves 0 separately. The two players will achieve 1 jointly, if they agree to form a coalition together.

Example 2: Pure Bargaining Game

In this game there are n players who can achieve, say, 1, provided they all participate; otherwise, they will achieve nothing. (Example 1 is a special case in which $n = 2$.)

$N = \{1, 2, ..., n\}$.

The coalition function of this game is:

$$v(S) = \begin{cases} 1 & S = N \\ 0 & S \neq N, S \subset N \end{cases}$$

In the above formula, S represents any coalition whose members are players in N, and v is defined for all subsets of N, that is, for all coalitions of N.

Example 3: Buyer-Seller Exchange Game

Player 1 has a "spare" house that he is willing to sell at a reasonable price. Player 2 is interested in buying the house. Under what conditions will the deal be made?

This deal can be made only if the seller sets a lower worth on the house than the buyer does, because *in a deal both parties must profit*; otherwise, the deal will not be made.[5]

Suppose the worth of the house to the seller is \$100,000; he will not sell the house for less than this amount (but he hopes to get a higher payment). Suppose the worth of the house to the buyer is \$150,000; he will not pay more than this amount (but he hopes to pay less). The buyer sets a higher worth on the house because he needs a place to live.

The deal can be made if the house is sold for, say, \$120,000; the seller will make a profit of \$20,000, and the buyer will buy a house for \$120,000, which in his opinion is worth \$150,000, so that he makes a profit of \$30,000.

When we translate this into a game we see that:

The seller owns a house: $v(1) = 100,000$

The buyer owns no house: $v(2) = 0$

$$\begin{aligned} v(1,2) &= 120,000 + \\ &+ (150,000 - 120,000) = \\ &= 150,000 \end{aligned}$$

[5] If both players set the same worth on the house it is immaterial who will get the house and who will remain with the money.

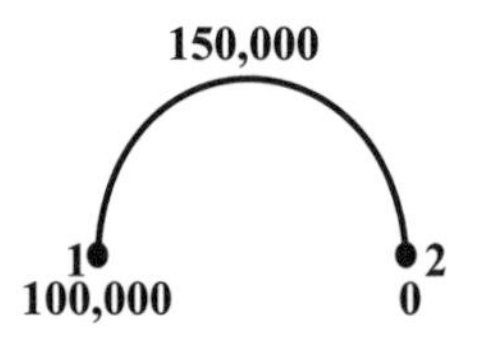

Explanation: Forming a coalition in this case means making a sale. After the sale, the seller has \$120,000 and the buyer has a house which he estimates to be worth \$150,000 and for which he paid a lesser amount; that is, he has \$150,000 minus the amount he paid for the house. This is the worth of the coalition.

Remark: The same function will be obtained even if the house is sold for a different price, p, such that $100,000 < p < 150,000$. Explain.

Example 4: Market Game with Two Sellers and One Buyer
There are three players in this game: two sellers and a buyer. Each seller has one good, say a DVD, for which he paid \$100, and offers to sell it. The buyer sets a worth of \$200 on the DVD. He is interested in paying the lowest possible price for the DVD, and, of course, he is unwilling to pay more than \$200.

Let us calculate the coalition function.

$N = \{1, 2, 3\}$; players 1 and 2 are the sellers and player 3 is the buyer.

Seller 1 has a DVD worth \$100 to him:	$v(1) = 100$
Seller 2 has a DVD worth \$100 to him:	$v(2) = 100$
Buyer 3 does not have a DVD:	$v(3) = 0$

Remark: The way this is written does not take into account money or other assets that the players may have. We could have added their worths but that would have had no effect on the solution that we shall now propose.

How shall we establish the worth of coalition $\{1, 3\}$? If 1 and 3 decide to form a coalition, it is reasonable to assume that player 1 will sell the DVD to player 3 for p dollars, when $100 < p < 200$. (Remember, the deal will be made only if both parties make a profit, so p must be greater than 100 for the seller and less than 200 for the buyer.) At this stage, player 1 has p dollars in cash and player 3 has a DVD which he estimates to be worth $200 and for which he paid a lesser amount; that is, he has $200 minus the amount that he paid for the DVD. Therefore, the worth of coalition $\{1,3\}$ is:

$$v(1,3) = p + (200 - p) = 200$$

The worth of coalition $\{2,3\}$ is established in exactly the same way as the worth of coalition $\{1,3\}$; that is, $v(2,3) = 200$, assuming players 2 and 3 decide to strike a deal.

The worth of coalition $\{1,2\}$ is $v(1,2) = 200$. This is a coalition of two sellers in which each seller has a DVD valued at $100 and so the coalition worth is $200.

How shall we establish the worth of coalition $\{1,2,3\}$? The formation of coalition $\{1,2,3\}$ denotes a procedure in which all players take part. Such a procedure may be carried out as follows. Seller 1 says to seller 2: "Quit the game and don't compete with me. In return, I'll pay you p dollars upon closing the deal." Then player 1 sells the DVD to buyer 3 for q dollars (of course, $q > p$). After the sale, the players are left with the following amounts:

Seller 1: the amount of $(q - p)$ dollars.
Seller 2: the amount of p dollars and a DVD worth $100 to him, for a total of $(100 + p)$.
Buyer 3: the amount of $(200 - q)$ dollars: a DVD worth $200 to him minus the amount q that he paid for the DVD.
Therefore,

$$v(1,2,3) = (q - p) + (100 + p) + (200 - q) = 300.$$

To summarize, the coalition function is:

$v(1) = 100$

$v(2) = 100$

$v(3) = 0$

$v(1, 2) = 200$

$v(1, 3) = 200$

$v(2, 3) = 200$

$v(1, 2, 3) = 300$

$v(\emptyset) = 0$

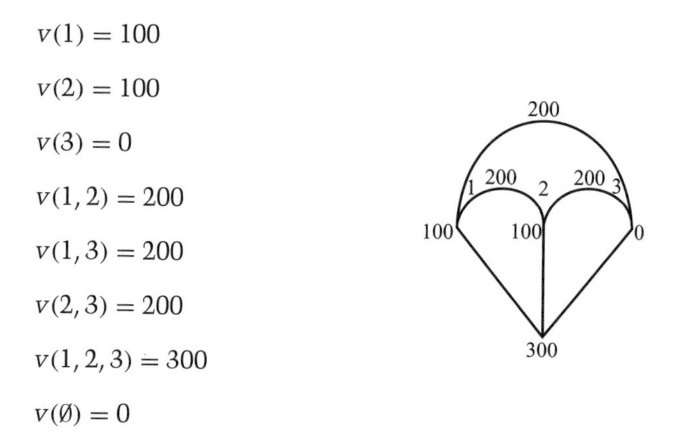

The coalition function describes the worth of all coalitions that can be formed under the above-mentioned conditions.

3.4 EXERCISES

1. A market game has two buyers and one seller. Seller 1 has a laptop whose worth for him is \$2000 and he offers it for sale. Players 2 and 3 are two potential buyers who do not own a laptop. The worth of the laptop is \$2,800 for one buyer (player 2) and \$3,000 for the other buyer (player 3). Each buyer is interested in paying the lowest possible price for the laptop and is of course unwilling to pay more than the worth of the laptop to him. Describe the game in coalition function form.

2. A glove-market game has three players: players 1 and 2 each have a left-hand glove and player 3 has a right-hand glove. The worth of a coalition is the amount that it will get for the gloves in its possession. Every pair of gloves (left and right) can be sold in the market for \$50. A single glove cannot be sold in the market. Describe the game in coalition function form.

3. A game v describes a four-player market in which there are two sellers and two buyers. Each seller has a good that he offers for sale and whose worth for him is \$100. Each buyer wants to buy a single good whose worth for him is \$150. The potential buyers do not have a good like the one offered for sale. Describe the game in coalition function form.

4. A glove-market game has five players: players 1, 2, and 3 each have a left-hand glove and players 4 and 5 each have a right-hand glove. The worth of a coalition is the amount that it will get for the gloves in its possession. Every pair of gloves (left and right) can be sold in the market for \$100. Describe the game in coalition function form.

3.5 ADDITIVE GAMES

A game $(N; v)$ is called *additive* if

$$v(S \cup T) = v(S) + v(T)$$

holds for every pair of disjoint coalitions S and T (namely, $S \cap T = \phi$).

A game is additive if every pair of disjoint coalitions in it are such that the members of any coalition obtain as a group exactly what they would have obtained had they been acting separately. Additive games are the most trivial games. In such games there is no incentive to form a coalition because no coalition provides a surplus over what its members can obtain on their own.

Example:

$v(1) = 5 \qquad v(2) = 10 \qquad v(3) = 15$

$v(1,2) = 15 \qquad v(1,3) = 20 \qquad v(2,3) = 25$

$v(1,2,3) = 30$

$v(\emptyset) = 0$

Verify the additivity.

3.6 SUPERADDITIVE GAMES

A game $(N; v)$ will be called *superadditive* if

$$v(S \cup T) \geq v(S) + v(T)$$

holds for every pair of disjoint coalitions S and T (namely, $S \cap T = \phi$). In other words, a superadditive game is a game in which every pair of disjoint coalitions can obtain jointly at least as much as they could have obtained if they had been acting separately. Or, to put it another way, it is advantageous to form large coalitions.

This requirement seems natural enough, since the large coalition is free to act as if it consisted of several coalitions.

The question is, how reasonable is it to assume superadditivity? The answer is that it depends on what real-life situation we want to describe. In most social and economic situations the superadditivity assumption is reasonable, but there are other situations in which it is not – say a situation in which for certain reasons (personal, political, racist, etc.) there are two groups that not only are not mutually beneficial, but one group harms the interests of the other group. In such a case, where unity is counterproductive, the superadditivity assumption is unreasonable. Another example is a situation in which an antitrust law prohibits small firms from merging into a large firm. In such a case a heavy fine might be imposed on the firms that merge in violation of the law.

In a superadditive game,

$$v(S \cup T) \geq v(S) + v(T)$$

holds for every pair of disjoint coalitions S and T; this inequality shows that for S and T there is an advantage to forming coalition $S \cup T$. In superadditive games it is advantageous to form the largest possible coalitions, so it may naturally be assumed that the coalition N will form and the comprehensive payoff to its members will be $v(N)$. This comprehensive payoff will have to be divided somehow among the members of coalition N – hence, in such cases, we can assume that the final outcome will be a *payoff vector* $x = (x_1, x_2, x_3, ..., x_n)$ that satisfies

$$x_1 + x_1 + ... + x_n = v(N).$$
$$x_1 \geq v(1), x_2 \geq v(2), ..., x_n \geq v(n).$$

Here, x_i is the payment to player i. The first line, called *group rationality*, or *efficiency*, indicates that the grand coalition formed and its worth was shared by all the players. The second line, called *individual rationality*, indicates that each player i will participate only if he gets at least what he could get by playing alone. We can sum up by saying that at least for superadditive games, the solution of the game, however it is obtained, should be in a set of payoff vectors.

Example:

$v(1) = 6$ $\quad v(2) = 18$ $\quad v(3) = 12$
$v(1,2) = 24$ $\quad v(1,3) = 20$ $\quad v(2,3) = 35$
$v(1,2,3) = 50$ $\quad v(\emptyset) = 0$

Verify the superadditivity.

3.7 MAJORITY GAMES

One of the election rules in a voting body defines which subsets of the voting body are big enough to pass a decision, and which do not meet this requirement. Those subsets that can pass a decision are called *winning coalitions*. The coalitions that are not big enough to pass a decision are called *losing coalitions*. Often, the winning coalitions are those that have a majority.

A voting body can be described as a coalition function game. To illustrate this, we assign a worth 1 to every winning coalition and a worth 0 to every losing coalition. In such games, the worth 1 is not a monetary payment but a more abstract worth, namely, the ability to pass decisions. The number 1 signifies that the winning coalition can achieve whatever it wants. The number 0 signifies that the losing coalition cannot guarantee itself anything.

Example 1: Three-player Majority Game

A class committee consists of three members. To pass a decision, the agreement of two members is sufficient. Of course, the agreement of all members is sufficient to pass a decision, while one committee member alone is a minority.

The coalition function in this case is:

$N = \{1, 2, 3\}$ $\qquad v(1) = v(2) = v(3) = 0 \qquad v(\emptyset) = 0$

$v(1, 2) = v(1, 3) = v(2, 3) = v(1, 2, 3) = 1$

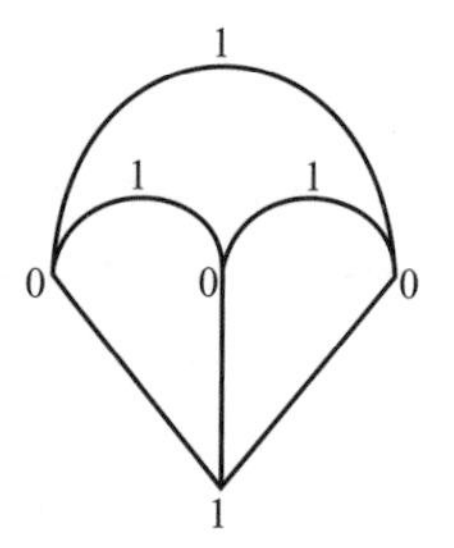

We can examine this game in two ways. One is to ask what coalitions can be formed and how they are likely to share their worth. If coalition {1, 2} is formed, the outcome will probably be (1/2, 1/2, 0). If coalition {1, 3} is formed, the outcome will probably be (1/2, 0, 1/2). If coalition {2, 3} is formed, the outcome will probably be (0, 1/2, 1/2).

The other way to examine this game is to ask what a judge is likely to decide if the class committee brings the game before him and asks him to propose a "fair" distribution. The judge's decision for this game will likely be that the grand coalition should form and share its worth equally. Thus, the outcome will be (1/3, 1/3, 1/3).

Example 2: Weighted Majority Game

In a voting body like a parliament, the players are parties and every party has a certain number of representatives. A certain country with three parties in its parliament obtained the following election results:

Party 1: 5 representatives
Party 2: 3 representatives
Party 3: 7 representatives

The number of representatives of party i is called the "weight" of party i. Let w denote the "weight" so that the weight of i is w_i. In our example, $w_1 = 5, w_2 = 3, w_3 = 7$.

These weights are especially justified in parliaments in which coalition disciplines are enforced, that is, in which it is mandatory that all members of a party vote the same way when important issues are considered. In such cases it is justified to call the party a player and the number of its representatives its weight.

In general, a weighted majority game, written $[q; w_1, w_2, ..., w_n]$, is a game $(N; v)$, where $w_1 \geq 0, w_2 \geq 0, ..., w_n \geq 0$ and

$$v(S) = 1 \text{ if the sum of the weight of } S \text{ is greater than or equal to } q$$
$$v(S) = 0 \text{ otherwise}$$

The number q is called the *quota* of the game. It is the number of weights a coalition must command in order to pass a decision.

In our example $v(1) = v(2) = v(3) = 0$, $v(1,2) = v(1,3) = = v(2,3) = 1$, and $v(1,2,3) = 1$.

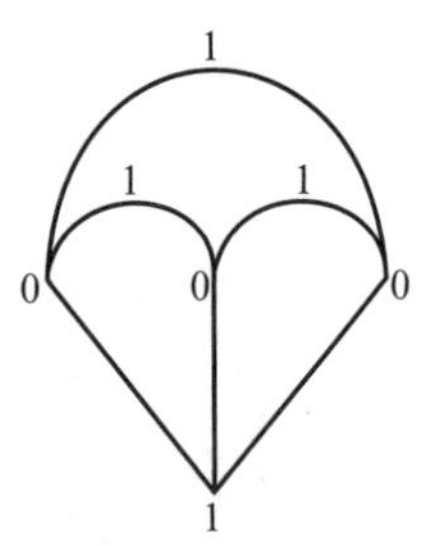

Thus, we can choose $q = 8$ and the game is $[8; 5, 3, 7]$. But we can also choose $q = 7\frac{1}{2}$ and the game is then $[7\frac{1}{2}; 5, 3, 7]$. If, on the other hand, one needs $\frac{2}{3}$ of the total sum of weights in order to pass a decision then $q = 10$ and the game is described as $[10; 3, 5, 7]$.

A priori, not all weighted majority games result from real-life elections. Take, for example, the weighted majority game $[8; 3, 4, 5, 6]$. Here both $S = \{1, 3\}$ and $T = \{2, 4\}$ are winning coalitions, as are many others. Can we say that both S and T control all the rulings in a parliament? Similarly, $[7; 2, 2, 2, 2, 2, 2]$ is a strange game, because both $\{1, 2, 3\}$ and $\{4, 5, 6\}$ are losing coalitions. We therefore limit ourselves to *strong weighted majority games*, where the quota is greater than half of the sum of all the weights. Strong weighted majority games satisfy the following properties:

1. An empty coalition is a losing coalition.
2. A coalition of all players is a winning coalition.
3. If S is a winning coalition, the coalition of players not in S is a losing coalition.
4. If S is a losing coalition, the coalition of players not in S is a winning coalition.

Exercise:
What is the coalition function of the game $[9; 1, 7, 9]$?

Answer:

Players 1 and 2 cannot pass a decision separately: $v(1) = v(2) = 0$
Player 3 can pass a decision separately: $v(3) = 1$
Players 1 and 2 together do not have enough votes (explain): $v(1, 2) = 0$

$$v(\emptyset) = 0 \qquad v(1, 3) = v(2, 3) = v(1, 2, 3) = 1$$

Or, expressed as a figure:

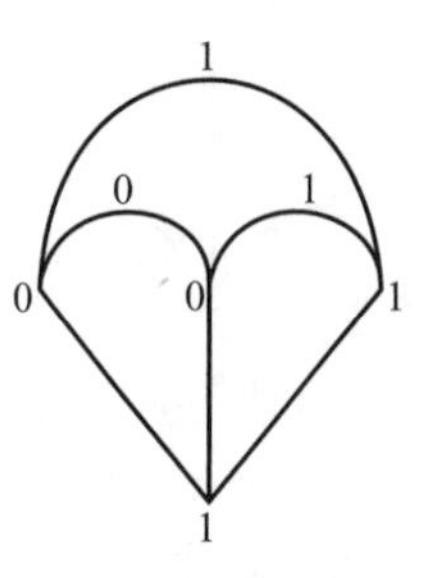

Player 3 in this game is called a *dictator* because his consent is necessary and sufficient to pass any decision.

3.8 EXERCISES

1. Write the coalition function of each of the following weighted majority games:

(1) $[3; 2, 1, 1]$	(5) $[9; 5, 5, 3, 4]$
(2) $[3; 2, 1, 1, 1]$	(6) $[61; 61, 19, 20, 20]$
(3) $[8; 6, 2, 7]$	(7) $[61; 50, 40, 30]$
(4) $[8; 5, 6, 4]$	(8) $[61; 35, 35, 35, 15]$

2. In the following games the required majority is not a simple majority, but a decisive majority. In each of these games a two-thirds majority is needed to pass a decision. Given this condition, write the coalition function of each of the following games:

(1) $N = \{1, 2, 3, 4, 5\}$	$w_1 = w_2 = 2$	$w_3 = w_4 = w_5 = 1$	
(2) $N = \{1, 2, 3, 4\}$	$w_1 = w_2 = 5$	$w_3 = 3$	$w_4 = 4$
(3) $N = \{1, 2, 3\}$	$w_1 = 50$	$w_2 = 40$	$w_3 = 30$
(4) $N = \{1, 2, 3, 4\}$	$w_1 = w_2 = w_3 = 35$	$w_4 = 15$	

3. Write the coalition function of the following game:
$N = \{1, 2, 3, 4\}$; a coalition is defined to be a winning coalition if and only if it contains player 1 and there are at least 3 players.

4. A player is called a *veto player* if he is in every winning coalition; that is, a coalition cannot be a winning coalition unless it contains him.

Are there veto players in the following weighted majority games? Support your answer!

(1) $[8; 1, 1, 1, 6]$
(2) $[7; 5, 5, 3]$
(3) $[61; 60, 20, 20, 20]$
(4) $[61; 59, 30, 21, 10]$
(5) $[10; 3, 3, 3, 9]$

3.9 SYMMETRIC PLAYERS

Example 1

Take the following game:

$N = \{1, 2, 3, 4\}$ $v(1) = v(2) = v(3) = v(4) = 0$ $v(\emptyset) = 0$
$v(1, 2) = 8$ $v(2, 3) = 9$
$v(1, 3) = 9$ $v(2, 4) = 15$
$v(1, 4) = 15$ $v(3, 4) = 5$
$v(1, 2, 3) = 20$ $v(1, 3, 4) = 50$
$v(1, 2, 4) = 30$ $v(2, 3, 4) = 50$
$v(1, 2, 3, 4) = 60$

The game can be seen to contain several pairs of coalitions of equal worth:

$v(1) = v(2) = 0$
$v(1, 3) = v(2, 3) = 9$
$v(1, 4) = v(2, 4) = 15$
$v(1, 3, 4) = v(2, 3, 4) = 50$

What characterizes each pair of coalitions of equal worth is that one coalition contains player 1 with additional players and the other coalition contains player 2 with the same additional players. Thus, in each coalition containing either player 1 or player 2, if we replace player 1 by player 2 or vice versa, the coalition worth will not change.

Question: Are the coalitions listed above the only ones that need to be checked?

Answer: All other coalitions either contain both players, in which case the replacement does not change the coalition, or, alternatively, they do not contain either of the two players, in which case there is no one to replace. (Verify!)

We can sum up by saying that in this game, in *every* coalition containing either player 1 or player 2, the coalition worth does not change when we replace one of these players with the other. In such a case we say that players 1 and 2 are *symmetric*.

Example 2

$N = \{1,2,3,4\}$ $v(1) = v(2) = 1$ $v(3) = v(4) = 0$

$v(1,2) = 7$ $v(2,3) = 7$

$v(1,3) = 10$ $v(2,4) = 6$

$v(1,4) = 5$ $v(3,4) = 5$

$v(1,2,3) = 20$ $v(1,3,4) = 25$

$v(1,2,4) = 30$ $v(2,3,4) = 40$

$v(1,2,3,4) = 70$ $v(\emptyset) = 0$

In this game, too, there are pairs of coalitions of equal worth:

$v(1) = v(2) = 1$

$v(3) = v(4) = 0$

$v(1,2) = v(2,3) = 7$

$v(1,4) = v(3,4) = 5$

Question: Can players 1 and 3 replace each other in every coalition that contains one of them?

Answer: No. Although player 1 can replace player 3 in coalition $\{2,3\}$ and player 3 can replace player 1 in coalition $\{1,2\}$, there are coalitions in which such a replacement changes the worth. For example, $v(1,2,4) \neq v(2,3,4)$, $v(1) \neq v(3)$.

Definition: Symmetric Players

Two players in a set N of players are called *symmetric players* in the game $(N; v)$ if they can replace each other in every coalition that contains one of them; that is, if one player replaces the other, the coalition worth does not change.

Formally, players i and j are called symmetric if for every S that does not contain i and j there exists

$$v(S \cup \{i\}) = v(S \cup \{j\}).$$

In particular, it should hold if $S = \emptyset$, in which case $v(i) = v(j)$.

3.10 EXERCISES

1. Following is a list of games. Check each of them to see whether it has symmetric players. If so, indicate who they are.

(1) $N = \{1, 2\} \quad v(N) = 1 \quad v(1) = v(2) = 0$

(2) $N = \{1, 2, ..., n\} \quad v(S) = \begin{cases} 1 & S = N \\ 0 & S \neq N,\ S \subset N \end{cases}$

(3) $N = \{1, 2, 3\} \quad v(1) = v(2) = 100 \quad v(3) = 0$
$v(1, 2) = v(1, 3) = v(2, 3) = 200$
$v(1, 2, 3) = 300 \quad v(\emptyset) = 0$

(4) $N = \{1, 2, 3\} \quad v(1) = v(2) = v(3) = 0 \quad v(\emptyset) = 0$
$v(1, 2) = v(1, 3) = v(2, 3) = v(1, 2, 3) = 1$

(5) $[3; 2, 1, 1, 1]$

(6) $N = \{1, 2, 3, 4\} \quad v(1) = v(2) = 2 \quad v(3) = v(4) = 0$
$v(1, 2) = 5 \quad v(2, 3) = 6$
$v(1, 3) = 7 \quad v(2, 4) = 5$
$v(1, 4) = 8 \quad v(3, 4) = 7$
$v(1, 2, 3) = 20 \quad v(2, 3, 4) = 15$
$v(1, 2, 4) = 10 \quad v(1, 3, 4) = 20$
$v(1, 2, 3, 4) = 30 \quad v(\emptyset) = 0$

2. Identify the symmetric players in the following weighted majority games:

(1) $[8; 6, 2, 7]$
(2) $[9; 5, 5, 3, 4]$
(3) $[61; 61, 19, 20, 20]$
(4) $[61; 50, 40, 30]$
(5) $[61; 35, 35, 35, 15]$

3.11 NULL PLAYERS

Take the majority game $[12; 1, 3, 7, 12]$. We shall write it as a coalition function:

$v(1) = v(2) = v(3) = 0 \quad v(4) = 1$
$v(1, 2) = 0 \quad v(2, 3) = 0$
$v(1, 3) = 0 \quad v(2, 4) = 1$
$v(1, 4) = 1 \quad v(3, 4) = 1$
$v(1, 2, 3) = 0 \quad v(1, 3, 4) = 1$
$v(1, 2, 4) = 1 \quad v(2, 3, 4) = 1$
$v(1, 2, 3, 4) = 1 \quad v(\emptyset) = 0$

We are interested in the worth of each coalition that contains player 1:

$v(1) = 0 \quad v(1, 2, 3) = 0$
$v(1, 2) = 0 \quad v(1, 2, 4) = 1$
$v(1, 3) = 0 \quad v(1, 3, 4) = 1$
$v(1, 4) = 1 \quad v(1, 2, 3, 4) = 1$

It can be seen that the participation of player 1 in any coalition does not change the coalition worth. Let us withdraw player 1 from every coalition that contains him and examine the worth of the resulting coalitions:

$v(\emptyset) = 0 \quad v(2, 3) = 0$
$v(2) = 0 \quad v(2, 4) = 1$
$v(3) = 0 \quad v(3, 4) = 1$
$v(4) = 1 \quad v(2, 3, 4) = 1$

When we compare this list to the previous one, we see that in this game the withdrawal of player 1 from a coalition that contains him, or, alternatively, the participation of player 1 in any coalition, does not change the coalition worth.

Formally, for every coalition S that does not contain player 1 $v(S \cup \{1\}) = v(S)$ is satisfied. A player whose presence or absence does not change the coalition worth is called a *null player*.

Definition: Null Player

A player in a set N of players is called a *null player* if he does not contribute anything through his participation in any coalition.

Formally, i is a null player if, for every S that does not contain i,

$$v(S \cup \{i\}) = v(S)$$

is satisfied. In particular, for $S = \emptyset$, we obtain $v(i) = 0$.

3.12 EXERCISES

1. (1) Identify the null players in the following game:

$$N = \{1, 2, 3, 4\} \quad v(1) = v(2) = v(3) = v(4) = 0 \quad v(\emptyset) = 0$$

$$\begin{array}{ll} v(1,2) = 0 & v(1,2,3) = 10 \\ v(1,3) = 0 & v(1,2,4) = 0 \\ v(1,4) = 0 & v(1,3,4) = 0 \\ v(2,3) = 10 & v(2,3,4) = 10 \\ v(2,4) = 0 & v(1,2,3,4) = 10 \\ v(3,4) = 0 & \end{array}$$

   (2) Does the answer change when $v(1, 2, 3, 4) = 20$ and for every other coalition S, $v(S)$ remains unchanged? (Support your answer!)

2. (1) Identify the null players and the symmetric players in the following game:

$$N = \{1, 2, 3, 4\} \quad v(1) = v(2) = 0 \quad v(3) = v(4) = 1 \quad v(\emptyset) = 0$$

$$\begin{array}{ll} v(1,2) = 0 & v(2,3) = 1 \\ v(1,3) = 1 & v(2,4) = 1 \\ v(1,4) = 1 & v(3,4) = 2 \\ v(1,2,3) = 1 & v(2,3,4) = 2 \\ v(1,2,4) = 1 & v(1,3,4) = 2 \\ v(1,2,3,4) = 2 & \end{array}$$

(2) Does the answer change when the value of $v(2)$ changes to $v(2) = 1$ and everything else remains unchanged? Support your answer!

3. Identify the null players in the following weighted majority games:

(1) $[61; 35, 35, 35, 15]$
(2) $[7; 5, 4, 3, 1]$
(3) $[10; 5, 5, 5, 2, 2]$

3.13 THE SUM OF GAMES

The mathematical model of a cooperative game is the pair $(N; v)$, where N is a set of players and v is the coalition function that assigns a real number to every coalition. An infinite number of games can be assigned to a set of players N by means of different coalition functions. Let us look, for example, at the following games assigned to the set $N = \{1, 2, 3\}$.

The game $(N; u)$ is:

$u(1) = u(2) = u(3) = 0 \quad u(\emptyset) = 0$
$u(1, 2) = 10 \qquad u(1, 3) = 20 \quad u(2, 3) = 30$
$u(1, 2, 3) = 40$

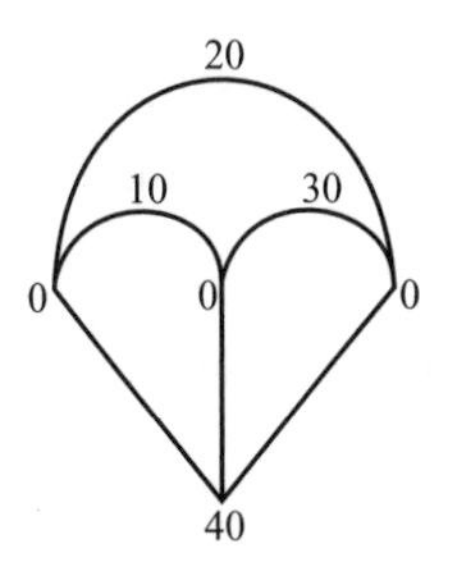

The game $(N; w)$ is:

$w(1) = 5 \qquad w(2) = 10 \qquad w(3) = 15 \qquad w(\emptyset) = 0$
$w(1, 2) = 20 \qquad w(1, 3) = 25 \quad w(2, 3) = 30$
$w(1, 2, 3) = 35$

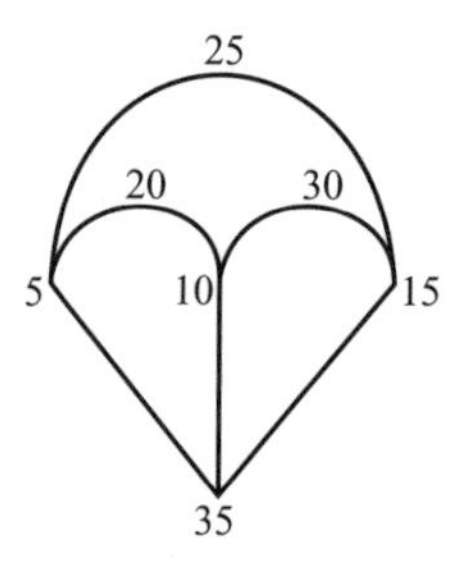

The game $(N; v)$ is:

$v(1) = 5$ $\quad v(2) = 10$ $\quad v(3) = 15$ $\quad v(\emptyset) = 0$

$v(1, 2) = 30$ $\quad v(1, 3) = 45$ $\quad v(2, 3) = 60$

$v(1, 2, 3) = 75$

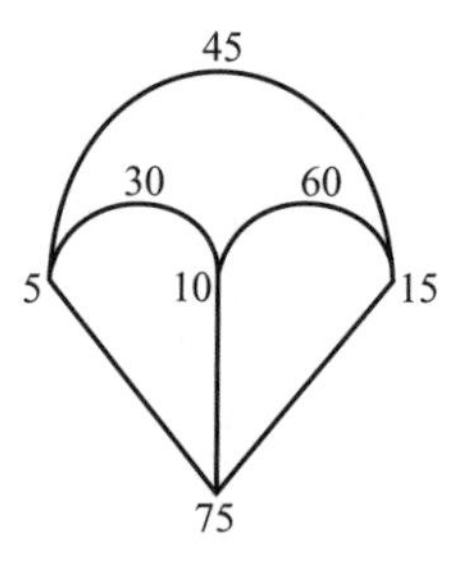

A look at the three games reveals that the game $(N; v)$ is the sum of the games $(N; u)$ and $(N; w)$, because the worth of every coalition in game v is the sum of its worths in games u and w.

For example, the worth of coalition $\{1, 2\}$ in game $(N; u)$ is 10 and its worth in game $(N; w)$ is 20, while its worth in the third game is 30.

This example leads to the following definition:

Definition: The Sum of Two Games

The game $(N; v)$ is called the *sum of two games* $(N; u)$ and $(N; w)$ if for every coalition S from the set of players N $(S \subseteq N)$

$$v(S) = u(S) + w(S)$$

This definition is also correct in the other direction; that is, given any game $(N; v)$, it is possible to split it into two games whose sum is the original game.

Example:

$N = \{1, 2, 3\}$

The coalition function is:

$v(1) = 10 \quad v(2) = 5 \quad v(3) = 15 \quad v(\emptyset) = 0$

$v(1,2) = 15 \quad v(1,3) = 30 \quad v(2,3) = 25$

$v(1,2,3) = 40$

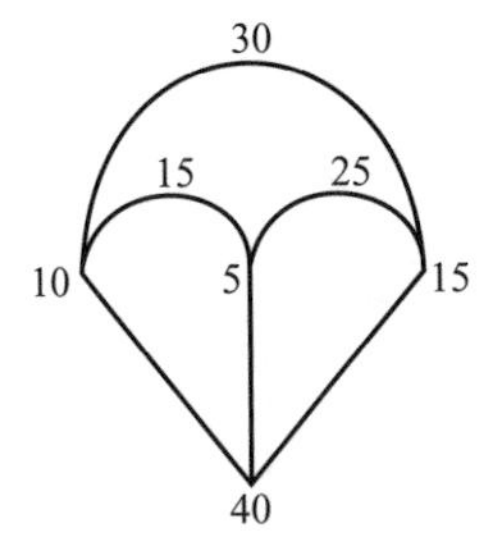

We shall examine some ways to split the game into two different games whose sum is the original game. For example, the game $(N; v)$ can be expressed as a sum of the two games $(N; w)$ and $(N; u)$, as in the following figure:

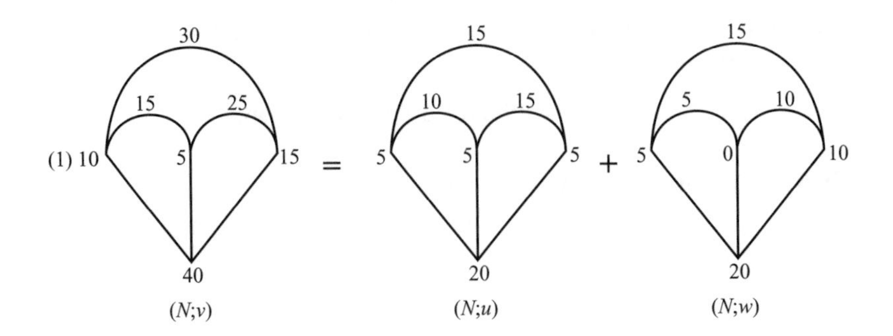

$w(1) = 5 \qquad u(1) = 5 \qquad \text{thus } w(1) + u(1) = 10 = v(1)$

$w(2,3) = 15 \quad u(2,3) = 10 \quad \text{thus } w(2,3) + u(2,3) = 25 = v(2,3)$

and so on.

Exercise:

Verify that in each of the given splits, for every coalition S, $v(S) = w(S) + u(S)$.

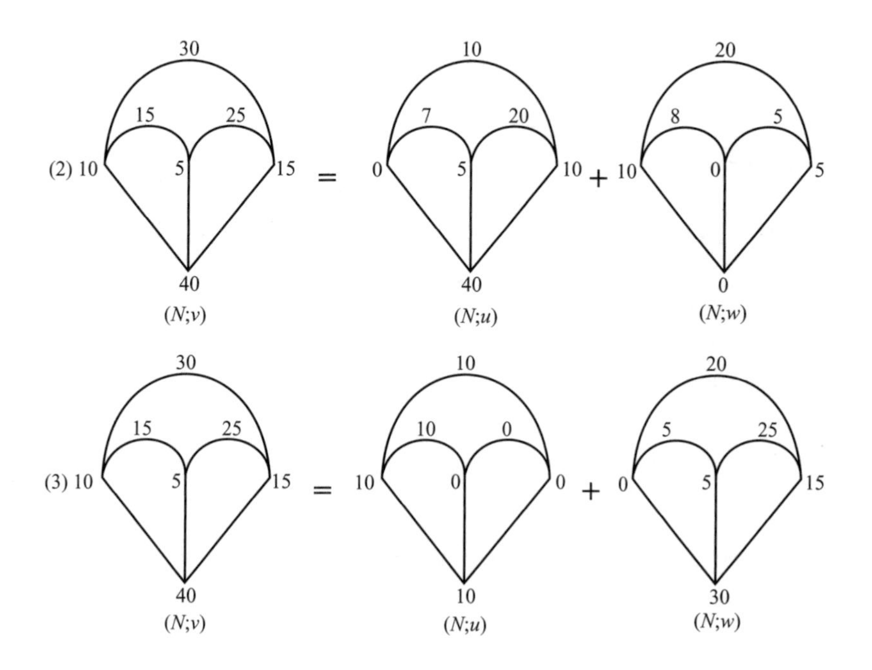

Exercise:

Find another way to split the original game into a sum of two games.

The concept of the sum of games enables us to create a new game from two given games when the same set of players is involved, and, conversely, to split a given game into two different games.

3.14 EXERCISES

1. A game is split into two different games below. Supply the missing values.

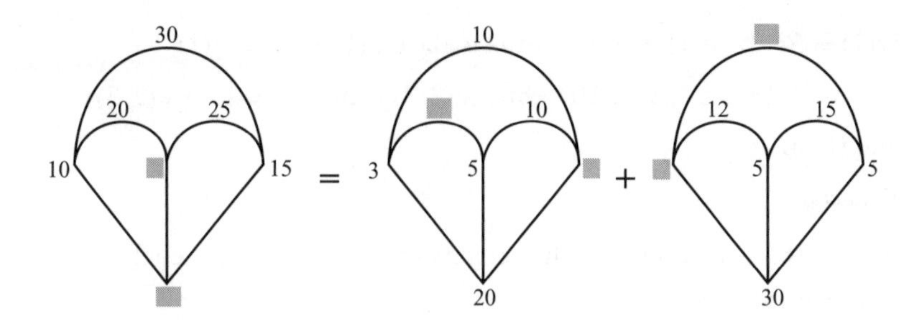

2. Take the two games $(N; v)$ and $(N; w)$. Write the coalition function of the sum of the games.

$N = \{1, 2, 3\}$

$v(1) = v(2) = v(3) = 0$

$v(1, 2) = 15$

$v(1, 3) = v(2, 3) = 20 \qquad v(\emptyset) = 0$

$v(1, 2, 3) = 30$

$w(1) = 5 \quad w(2) = 0 \quad w(3) = 5$

$w(1, 2) = 5 \quad w(1, 3) = 15 \quad w(2, 3) = 10$

$w(1, 2, 3) = 40 \quad w(\emptyset) = 0$

3. Take the game $(N; v)$:

$N = \{1, 2, 3\} \quad v(1) = v(2) = 5 \quad v(3) = 10 \quad v(\emptyset) = 0$

$v(1, 2) = 15 \quad v(1, 3) = 20 \quad v(2, 3) = 20$

$v(1, 2, 3) = 40$

The game is split into the sum of the two games below so that in one game player 1 is a null player.

(1) Supply the missing values of the second game.

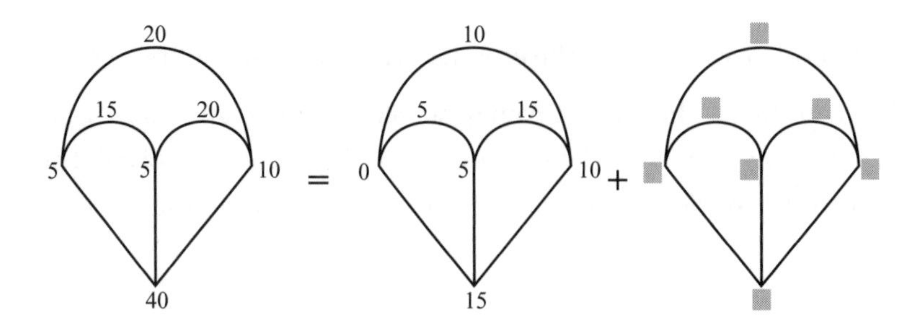

(2) Is there a null player in the second game too?

4. Take the game $(N; v)$:

$N = \{1, 2, 3\}$ $\quad v(1) = 10$ $\quad v(2)=5$ $\quad v(3) = 20$ $\quad v(\emptyset) = 0$
$v(1, 2) = 20$ $\quad v(1, 3) = 30$ $\quad v(2, 3) = 40$
$v(1, 2, 3) = 60$

(1) Split the game into the sum of two games, so that in one game players 2 and 3 are null players.
(2) Are there null players in the second game too?

5. Take the game $(N; v)$:

$N = \{1, 2, 3\}$ $\quad v(1) = 15$ $\quad v(2) = 15$ $\quad v(3) = 20$ $\quad v(\emptyset) = 0$
$v(1, 2) = 30$ $\quad v(1, 3) = 40$ $\quad v(2, 3) = 35$
$v(1, 2, 3) = 70$

The game is split into the two games below, so that in one game player 3 is a null player and players 1 and 2 are symmetric.

(1) Supply the missing values of the second game.

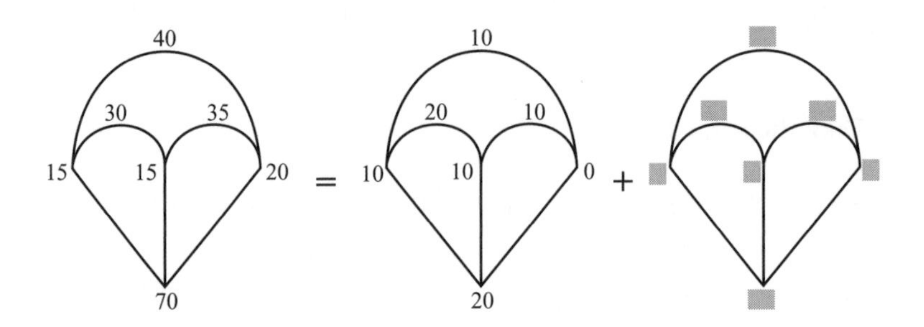

(2) Are there null players and/or symmetric players in the second game too?

6. Take the game $(N; v)$:

$N = \{1, 2, 3\}$ $\quad v(1) = 6$ $\quad v(2) = 6$ $\quad v(3) = 12$ $\quad v(\emptyset) = 0$
$v(1, 2) = 12$ $\quad v(1, 3) = 18$ $\quad v(2, 3) = 24$
$v(1, 2, 3) = 30$

(1) Split the game, so that in one game players 1 and 3 are symmetric players and player 2 is a null player.
(2) Are there null players and/or symmetric players in the second game too?

7. Take the game $(N; v)$:

$$N = \{1,2,3\} \quad v(1) = 9 \quad v(2) = 6 \quad v(3) = 12 \quad v(\emptyset) = 0$$
$$v(1,2) = 15 \quad v(1,3) = 24 \quad v(2,3) = 18$$
$$v(1,2,3) = 60$$

The game is split into the three games below, whose sum is the original game. Supply the missing values.

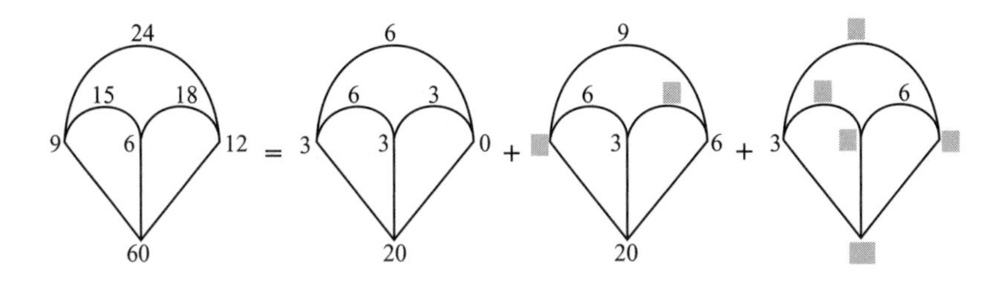

8. Take the game $(N; v)$:

$$N = \{1,2,3\} \quad v(1) = 5 \quad v(2) = 10 \quad v(3) = 15 \quad v(\emptyset) = 0$$
$$v(1,2) = 15 \quad v(1,3) = 25 \quad v(2,3) = 30$$
$$v(1,2,3) = 50$$

Split the game into three games whose sum is the original game.

3.15 THE SHAPLEY VALUE

For a game $(N; v)$, one can ask several questions: Given that certain coalitions formed, how should the players share the coalition worth among themselves? How will an unbiased judge or an arbitrator divide the proceeds if the players seek his advice? Each theory that attempts to answer any of these questions is called a *solution concept*. In this chapter we shall deal with one solution concept[6] called "the Shapley value." One of its aims is indeed to lay down rules that will enable an unbiased judge to suggest an allocation of $v(N)$ among the players.

We shall now present a simple system of four rules (axioms) and we shall see that this system of axioms offers the judge an opportunity to decide how to divide $v(N)$ fairly among the players in any given situation (in any game). We shall see that these axioms determine a unique way to divide $v(N)$ in every game.

The axioms presented below were first formulated in 1953 by Lloyd Shapley, who showed that indeed they dictate to the judge how to decide in every case. The division of payoffs according to this decision is called the *Shapley value.*

Example 1

$$N = \{1, 2, 3, 4\} \quad v(N) = 80$$
$$v(S) = 0, S \neq N$$

All players in this game are symmetric. (Verify!) The judge will duly take note of this and suggest an equal division of $v(N)$ among the players. Hence the outcome will be $(20, 20, 20, 20)$.

This division is obtained from the following axioms:

Axiom 1

The total amount $v(N)$ is divided among all the players.

This axiom is called the *efficiency axiom,* because in many games $v(N)$ is the largest amount of money the players can get by splitting N into several coalitions. The axiom is reasonable, because the players want to divide among themselves everything they can achieve in the game when they unite, namely, exactly $v(N)$.

Axiom 2

Symmetric players get equal payoffs.

This axiom is called the *symmetry axiom* for obvious reasons. It is reasonable, because we seek a "fair" division that will be acceptable

[6] Shapley, L. S. 1953. "A value for n-person games," in Kuhn, H. and Tucker, A. W. (eds.), *Contributions to the theory of games II,* Annals of Mathematics Studies 28. Princeton: Princeton University Press, pp. 307–17

to all players, thus avoiding unnecessary bargaining. A division of payoffs of this sort, which does not discriminate between a player and his equal, will be acceptable to everyone.

Example 2

$N = \{1,2,3\}$ $v(1) = v(2) = v(3) = 0$ $v(\emptyset) = 0$

$v(1,2) = 30$ $v(1,3) = 0$ $v(2,3) = 0$

$v(1,2,3) = 30$

In this game, players 1 and 2 are symmetric players and player 3 is a null player (verify it!). It is reasonable to assume that player 3 will get nothing in the division of payoffs. Indeed, since he contributes nothing, it is reasonable that he should get nothing. Hence the division of payoffs in this game is $(15, 15, 0)$.

We now present a third axiom.

Axiom 3

The payoff to a null player is zero.

The axiom is called the *null player axiom*. It is reasonable, because it is to be expected that a player who contributes nothing should get nothing.

Example 3

$N = \{1,2\}$ $v(1) = 30$ $v(2) = 20$ $v(1,2) = 80$ $v(\emptyset) = 0$

We shall split the game into a sum of three games.

$u(1) = 30$ $u(2) = 0$ $u(1,2) = 30$

$w(1) = 0$ $w(2) = 20$ $w(1,2) = 20$

$\tau(1) = 0$ $\tau(2) = 0$ $\tau(1,2) = 30$

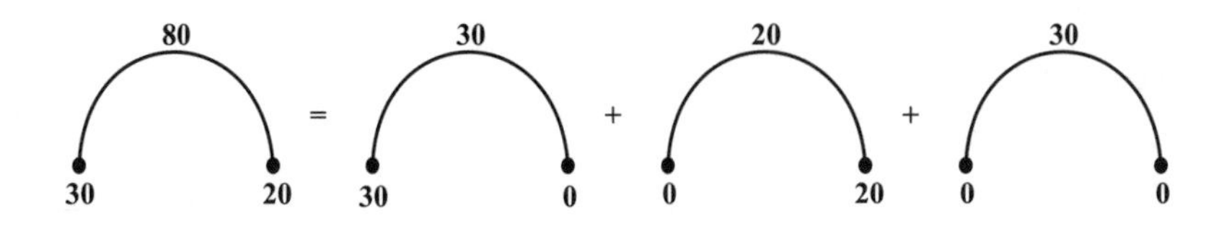

In game $(N; u)$ player 2 is a null player, in game $(N; w)$ player 1 is a null player, while in game $(N; \tau)$ players 1 and 2 are symmetric players. (Verify it!)

Based on the axioms presented above, the division of payoffs for each of the games is:

$(N; u) - (30, 0)$

$(N; w) - (0, 20)$

$(N; \tau) - (15, 15)$

If we add up these sums, we get $(45, 35)$, and that, we believe, is the division that the judge should propose.

We have based the judge's decision on the following axiom:

Axiom 4

If we split the original game into a sum of individual games, the division of payoffs among the players in the original game should be the sum of divisions obtained in the individual games.

We justify this axiom as follows. Let us imagine a situation where the set of players plays separately in each individual game; in this case the players will ultimately get the sum of payoffs that they get in each individual game.

On the other hand, if we consider the collection of games as a single game, we get a game whose coalition function is the sum of the coalition functions of the individual games, because every coalition can guarantee its worth in each individual game and, ultimately, it guarantees the sum of these worths.

Example:

Here are two games:

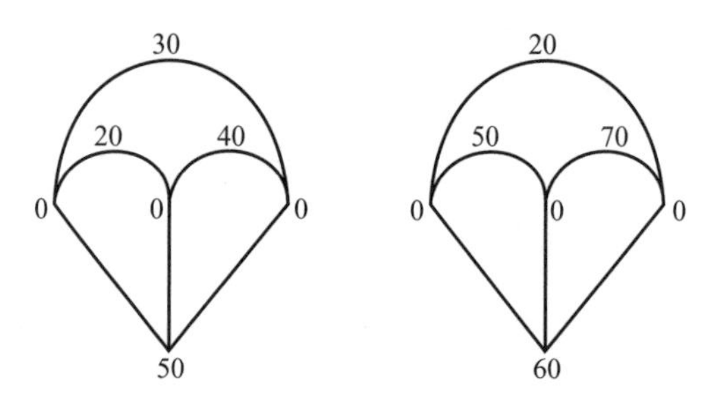

Let's look at coalition $\{1, 2\}$. In the first game this coalition can guarantee itself 20, if the players decide to form such a coalition. In the second game the coalition will be able to guarantee itself 50, if such a coalition is formed. Ultimately, the coalition can guarantee itself 70, if such a coalition is formed in both games. Similarly, we can add up the worth of the remaining coalitions, and the game obtained is:

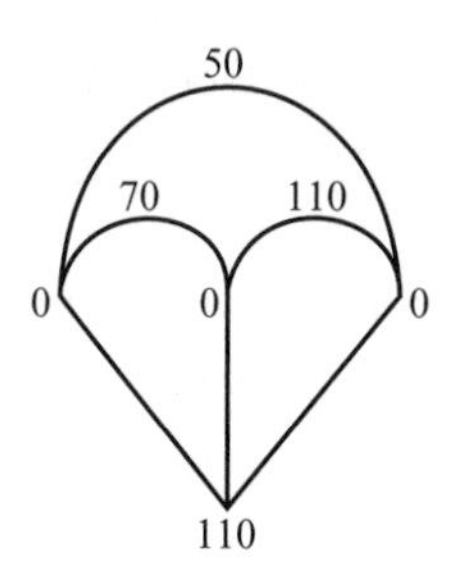

Let us reformulate Shapley's axioms concisely:

Efficiency axiom: $v(N)$ is completely divided among all the players.
Symmetry axiom: If there are symmetric players in a game, they get equal payoffs.
Null player axiom: If there is a null player in a game, he gets zero.
Additivity axiom: If $(N; v) = (N; w) + (N; u)$, the value of the sum of the two games equals the sum of the values of the two games.

Shapley, who formulated these axioms, proved that they establish a *unique value* for every game. While proving the Shapley theorem is beyond the scope of this book, we shall see how to find the value of a game in the examples below.

Example:

$N = \{1, 2, 3\}$ $\quad v(1) = 6 \quad v(2) = 12 \quad v(3) = 18$

$v(1, 2) = 30 \quad v(1, 3) = 60 \quad v(2, 3) = 90$

$v(1, 2, 3) = 120 \quad v(\emptyset) = 0$

There are no symmetric players and no null players in this game. The game can be split into a sum of very simple games so that there will be only null players and/or symmetric players in every game. We can calculate the Shapley value of these games by the axioms above, and then, by the additivity axiom, we can calculate the Shapley value of the original game.

We now present one way of splitting the original game into games in which there are only null players and/or symmetric players.

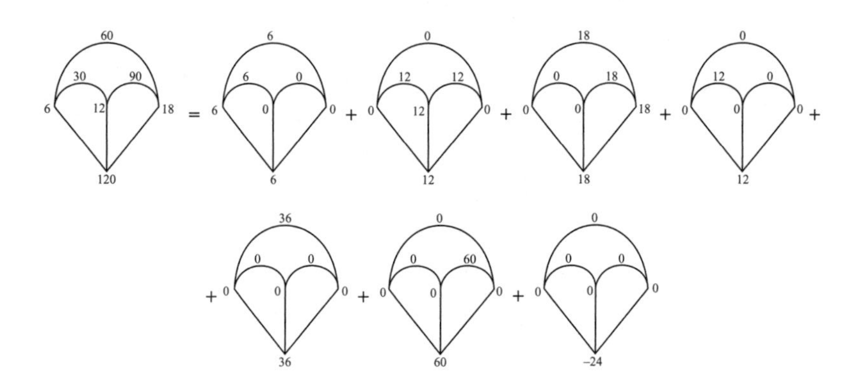

Verify that the sum of all these games is indeed the original game.

How is this split achieved?

First, we split the original game into two games.

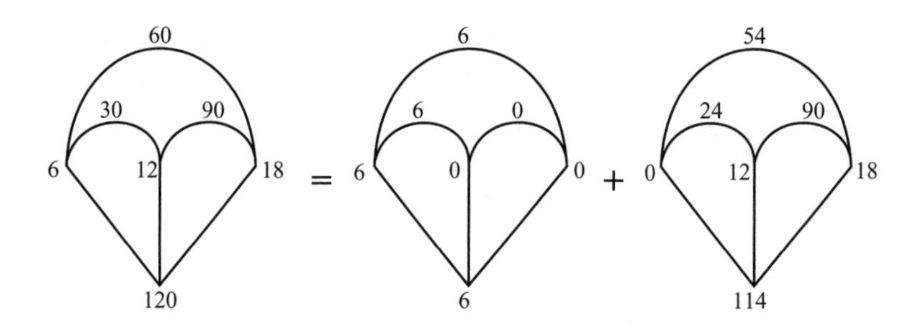

We then split the right-hand game into two games.

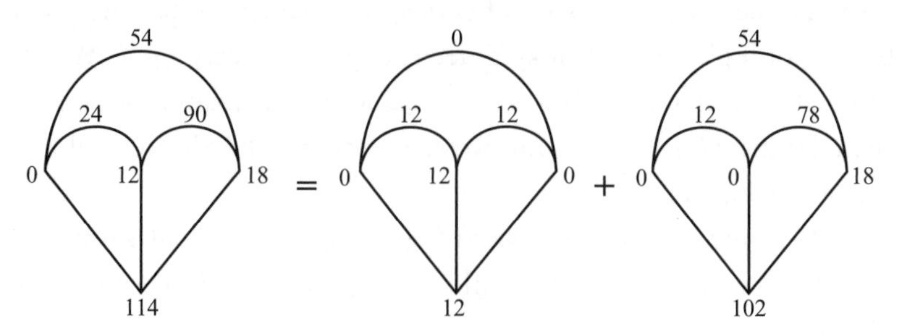

We can again split the right-hand game into two games.

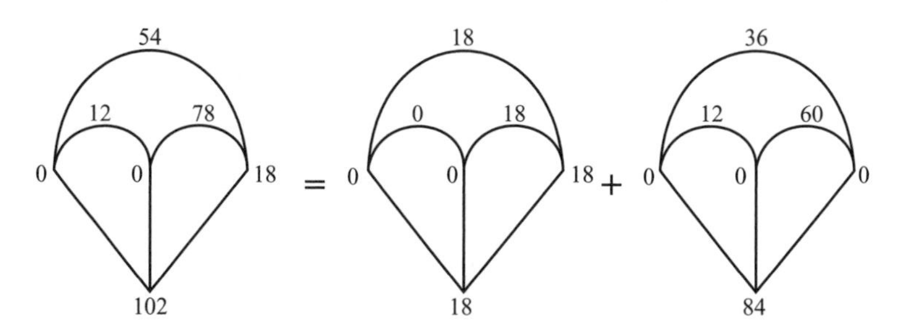

And so on.

Exercise:
Split the right-hand game into two games and keep splitting the games obtained until you have seven games, as in the example above.

Explanation: The split presented above is not random. At every stage we split the game into two games, one of which has a special property; namely, it contains a coalition S such that $v(T) = v(S)$ whenever T contains S and $v(T) = 0$ for every other coalition. Such a game is called a *carrier game* and the coalition S is called its *carrier*. Formally,

Definition: A *carrier game* $(N; v)$ is a game in which there is a coalition S, called the *carrier* of the game, such that

$$\begin{aligned} v(T) &= v(S), \quad \text{whenever } S \subseteq T \\ v(T) &= 0, \qquad \text{otherwise.} \end{aligned}$$

In the first three games the carriers are $\{1\}$, $\{2\}$, and $\{3\}$.

Let's move on to the fourth game. The carrier coalition in it is $\{1, 2\}$. The fourth game is obtained by splitting the following game:

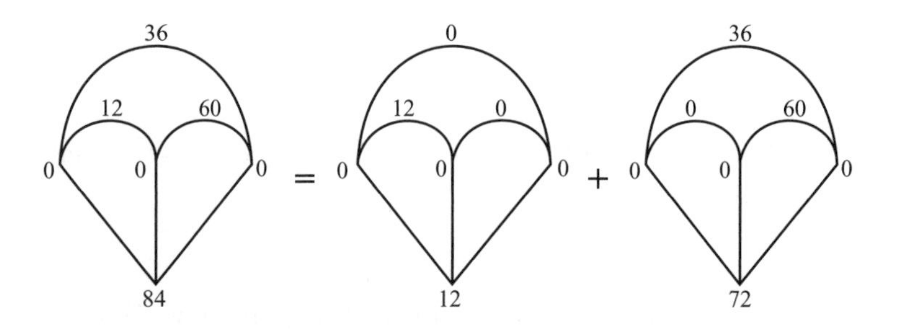

In this game the worth of coalition $\{1, 2\}$ is 12 $(v(1, 2) = 12)$. Hence the worth of the carrier coalition in the fourth game is 12. The remaining calculations are carried out similarly. Verify it!

Why did we split the game in this way? The Shapley value of every component of the split can be calculated by the axioms, because the carrier coalition players are symmetric players and the remaining players are null players. For example, in the following game:

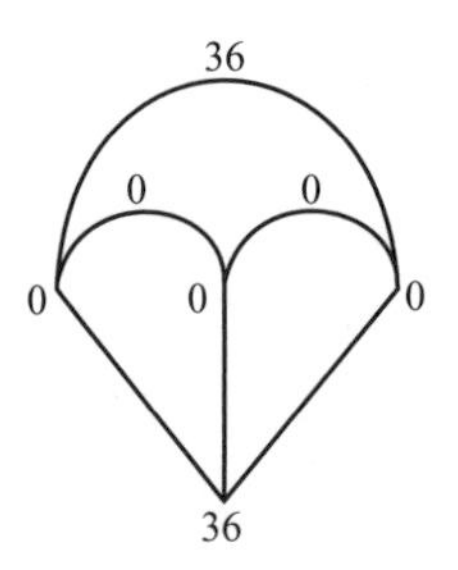

Players 1 and 3 are symmetric players and player 2 is a null player. Hence the Shapley value of the game is $(18, 0, 18)$.

If we examine the diagram of the split, we can calculate the following values:

$(6, 0, 0)$
$(0, 12, 0)$
$(0, 0, 18)$
$(6, 6, 0)$
$(18, 0, 18)$
$(0, 30, 30)$
$\underline{(-8, -8, -8)}$
$(22, 40, 58)$

Ultimately, by the additivity axiom, the value obtained is the Shapley value of the original game.

Exercise:

We now present another way of splitting the same game. Here again there are only null players and symmetric players in every component.

(1) Verify that the sum of games is the original game.

(2) Calculate the Shapley values of all the games and show that an identical value is obtained for the original game.

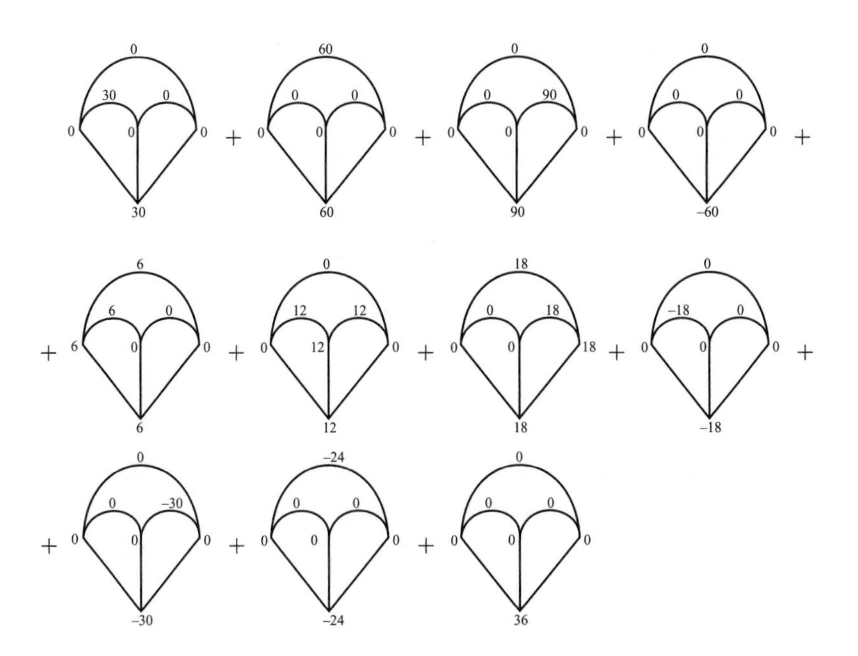

The result obtained – identical value – derives from the theorem proved by Shapley, according to which the aforementioned system of axioms establishes a unique value.

3.16 EXERCISES

1. Calculate the Shapley value of the two-player bargaining game.

$$N = \{1,2\} \quad v(1) = v(2) = 0 \quad v(1,2) = 1 \quad v(\emptyset) = 0$$

2. Calculate the Shapley value of the three-player majority game.

$$N = \{1,2,3\} \quad v(1) = v(2) = v(3) = 0 \;\; v(\emptyset) = 0$$
$$v(1,2) = v(1,3) = v(2,3) = v(1,2,3) = 1$$

3. Calculate the Shapley value of the following weighted majority games:

(1) $[61; 61, 19, 20, 20]$
(2) $[8; 5, 6, 4]$
(3) $[61; 35, 35, 35, 15]$

4. Calculate the Shapley value of the following game:

$$N = \{1,2,3\} \quad v(1) = v(2) = 100 \quad v(3) = 0 \quad v(\emptyset) = 0$$
$$v(1,2) = v(1,3) = v(2,3) = 200$$
$$v(1,2,3) = 300$$

5. Calculate the Shapley value of the following game:

$$N = \{1,2,3\} \quad v(1) = 6 \quad v(2) = 12 \quad v(3) = 18$$
$$v(1,2) = 12 \quad v(1,3) = 12 \quad v(2,3) = 24$$
$$v(1,2,3) = 48 \quad v(\emptyset) = 0$$

3.17 DISSOLVING A PARTNERSHIP

In this section we shall discuss the following real-life conflict[7]: we shall consider a joint enterprise whose owners decide to sell and divide

[7] Maschler, M. 1982. "The worth of a cooperative enterprise to each member," in Diestler, M., Furst, E. and Schwodiauer, G. (eds.), *Games, economic dynamics and time series analysis*. New York: Springer, pp. 67–73

its worth among themselves. How ought they to share the proceeds of the sale? One often-used method is to divide the proceeds in proportion to the investments of the partners. Variants of this method are employed in various ways in different firms. Another method is based on a principle often employed in "enterprises" consisting of a husband and wife who decide to seek a divorce settlement. According to this method, the wife takes back everything she brought to the marriage, the husband takes back everything he brought to the marriage, and the rest is divided equally between them. The question is whether this procedure can be generalized to a case of more than two partners.

To generalize a "marriage enterprise," let $N = \{1, 2, ..., n\}$ be a set of partners in an enterprise with joint ownership of its assets. As long as the enterprise operates, these assets serve its needs and all profits are divided among the partners in a way that need not concern us here (we do not rule out the possibility that the profits are divided in accordance with the arguments in the sequel). Now, the partners decide to dissolve and we are concerned with the division of assets among them. The situation is that some players contributed part of their assets to the enterprise, which they are entitled to claim back when the enterprise is dissolved. Also, some groups of partners have assets that they brought to the enterprise and ought to claim back and, of course, some assets were acquired by all the partners and should be distributed to all the partners. Of course, there are also claims against the assets that developed during the years when the enterprise was active and the partners are responsible for them. How can we take into account all of these factors when the enterprise is dissolved and sold in the market?

We shall now discuss a special case. A joint enterprise of three partners – 1, 2, and 3 – consists of a garage, a gas station, an auto accessories store, a restaurant, and an auto parts store. The history of the enterprise is as follows: the garage owner, 1, formed a partnership with the gas station owner, 2. The two worked together for a while and since their business prospered they bought an auto accessories store, run by a worker hired to serve the customers. Later, their neighbor, the restaurant owner, 3, joined the partnership, and the three opened an auto parts store. Unfortunately, differences of opinion arose among the partners, making it impossible for them to continue the partnership, and they agreed to sell everything and dissolve the partnership. Notice that 1 would remain the garage owner, 2 the gas station owner; similarly, 1 and 2 own the auto accessories store, and so on.

$\{1\} \to \{\text{garage}\}$ $\quad \{2\} \to \{\text{gas station}\}$

$\{3\} \to \{\text{restaurant}\}$ $\quad \{1, 2\} \to \{\text{auto accessories store}\}$

$\{1, 3\} \to \emptyset$ $\quad \{2, 3\} \to \emptyset$ $\quad \{1, 2, 3\} \to \{\text{auto parts store}\}$

We can clearly see that each individual partner has an asset and that partners 1 and 2 together have a joint asset. Partners 1 and 3 have no joint asset, partners 2 and 3 have no joint asset, partners 1, 2, and 3 have a joint asset and thus they have a joint enterprise composed of several assets.

We shall now model the situation as a game $(N; v)$, where $v(S)$ is the price that can be obtained in the market for *the set of all assets that belong to all subsets of* S*, including* S *itself*.

The worth is measured in fixed monetary units. We have chosen to base the definition of $v(S)$ on the set of all assets that belong to all subsets of S, rather than on the set of all assets that belong to just S. It is possible, for example, that a higher price will be obtained if all assets of all subsets of S are sold together as a package and not as separate items. It follows that when $v(S)$ is claimed by S, then it is necessary to decide on its division among the members of S. This division must take into account the claims $v(R)$ for $R \subseteq S$.

The coalition function in the case at hand could be:

$v(1) = 30 \quad v(2) = 12 \quad v(3) = 6 \quad v(1, 2) = 36$

$v(1, 3) = 36 \quad v(2, 3) = 30 \quad v(1, 2, 3) = 90 \quad v(\emptyset) = 0$

Explanation: The garage can be sold in the market for \$30,000. Similarly, the gas station and the restaurant can be sold separately for \$12,000 and \$6,000, respectively. The worth of coalition $\{1, 2\}$ is the worth of a package consisting of the garage, the gas station, and the auto accessories store. Note that $v(1, 2) < v(1) + v(2)$ even though this package contains the separate assets of 1 and 2. The reason is that there is a large mortgage on the auto accessories store for which both 1 and 2 are responsible.

A higher price can be obtained for the gas station and the restaurant if those assets are sold as a package and not as separate items. This is why

$$v(2,3) > v(2) + v(3).$$

We can provide similar explanations for the other coalition worths.

In order to describe the procedure for dividing the value of the joint enterprise among the partners in a way that is similar to the divorce settlement, we start with a numerical example. As mentioned, the partners want to liquidate the joint enterprise and divide the value, 90, among themselves. We shall attempt to solve this as follows.

Player 3, let us say, claims his share, namely, the value of the restaurant; the restaurant is sold and he gets 6. We need to subtract 6 from every coalition containing player 3, because every such coalition contains the asset of the restaurant. Thus, we get a new game:

$$u(1) = 30 \quad u(2) = 12 \quad u(3) = 0 \quad u(1,2) = 36$$
$$u(1,3) = 30 \quad u(2,3) = 24 \quad u(1,2,3) = 84$$

Suppose coalition $\{1,2\}$ claims its share, namely, the worth of all assets belonging to $\{1,2\}$ and its subsets. The coalition gets 36, which is divided equally between players 1 and 2. We subtract 36 from every coalition containing $\{1,2\}$, and thus we get a new game:

$$w(1) = 30 \quad w(2) = 12 \quad w(3) = 0 \quad w(1,2) = 0$$
$$w(1,3) = 30 \quad w(2,3) = 24 \quad w(1,2,3) = 48$$

We subtract 36 from $v(1,2,3)$ as well, because all assets of coalition $\{1,2\}$ and its subsets are included in the worth of $\{1,2,3\}$.

Suppose 1 now claims his share, 30, which is the worth of the garage. But the garage was sold already at the previous stage, when $\{1,2\}$ claimed its share, because $\{1\} \subset \{1,2\}$. We shall solve this problem as follows: we give player 1 the value of the garage, 30, and we place a debt of 30 on every coalition containing 1.We get the following game:

$p(1) = 0 \quad p(2) = 12 \quad p(3) = 0 \quad p(1,2) = -30$
$p(1,3) = 0 \quad p(2,3) = 24 \quad p(1,2,3) = 18$

And so on.

The following table presents a possible liquidation that leads to a final division of payoffs of $(39, 27, 24)$.

		coalition worth							division		
claimant	game	{1}	{2}	{3}	{1,2}	{1,3}	{2,3}	{1,2,3}	1	2	3
{3}	v	30	12	6	36	36	30	90	0	0	0
{1,2}	u	30	12	0	36	30	24	84	0	0	6
{1}	w	30	12	0	0	30	24	48	18	18	0
{2}	p	0	12	0	–30	0	24	18	30	0	0
{2,3}		0	0	0	–42	0	12	6	0	12	0
{1,2,3}		0	0	0	–42	0	0	–6	0	6	6
{1,2}		0	0	0	–42	0	0	0	–2	–2	–2
{1,2,3}		0	0	0	0	0	0	42	–21	–21	0
		0	0	0	0	0	0	0	14	14	14
									39	27	24

The procedure presented in the table above thus generalizes the divorce settlement, because every partner takes back what he brought to the partnership, while the rest is divided equally among all partners. Yet the procedure raises a few questions: Does every sequence of events (claims) lead to the same division of payoffs? Is the procedure necessarily finite? Is a division of $v(N)$ always obtained? We shall answer these questions shortly, but first we shall present another sequence of events – the shortest possible sequence – which is obtained when first all the one-player coalitions claim their worth, then all the two-player coalitions claim their worth, and finally the grand coalition claims its worth.

This sequence of events, originally offered by John C. Harsanyi,[8] is presented in the following table:

		coalition worth							division		
claimant	game	{1}	{2}	{3}	{1,2}	{1,3}	{2,3}	{1,2,3}	1	2	3
{1}	v	30	12	6	36	36	30	90	0	0	0
{2}	u	0	12	6	6	6	30	60	30	0	0
{3}	w	0	0	6	–6	6	18	48	0	12	0
{1,2}	p	0	0	0	–6	0	12	42	0	0	6
{2,3}		0	0	0	0	0	12	48	–3	–3	0
{1,2,3}		0	0	0	0	0	0	36	0	6	6
		0	0	0	0	0	0	0	12	12	12
									39	27	24

The fact that the final division of payoffs is exactly the same as before is an indication that this procedure is interesting. Later we shall see that the final division of payoffs is the Shapley value of the original game.

In words, the rule for dissolving a partnership is:
Split the original game into a sequence of games, where the divisions of payoffs are obtained as follows: in any game a certain coalition whose worth according to the coalition function is not zero claims its worth, which is divided equally among the members of that coalition.

The next game in the sequence is obtained as follows: subtract the worth of the claiming coalition from the worth of every coalition containing the claiming coalition and if a coalition does not contain the claiming coalition, then its worth according to the coalition function does not change, and so on.

The procedure terminates when the worth of every coalition is zero, and the final division of payoffs is the amount of money that accumulates over all the stages.

We shall prove that this procedure is valid by the following two theorems.

[8] Harasanyi, J. C. 1959. "A bargaining model for the cooperative n-person game," in Tucker, A. W. and Luce, R. D. (eds), *Contributions to the theory of games IV*, Annals of Mathematics Studies 40. Princeton: Princeton University Press, pp. 325–55

Theorem:

The procedure for dissolving a partnership terminates in a finite number of steps.

Proof:

When a coalition claims its worth, its worth in the next game is zero. The claiming coalition can again obtain a worth that is not zero only if one of its non-zero subcoalitions (i.e., whose worth is not zero) claims its worth. Suppose the procedure is infinite; in this case there exists at least one coalition that claims its worth infinitely many times. Accordingly, there exists a minimal coalition S that claims its worth infinitely many times as well. But this can happen only if one of its non-zero subcoalitions claims its worth infinitely many times, which contradicts the minimality of S. The contradiction shows that the assumption that the procedure is infinite is false; that is, a procedure for dissolving a partnership must terminate in a finite number of steps.

Theorem:

No matter what sequence is chosen for the procedure for dissolving a partnership, the final division of payoffs is the Shapley value of the game $(N; v)$.

The proof of the theorem requires the use of many symbols, so we shall not undertake it here. Instead, we shall clarify the idea of the proof by way of example.

Consider the example at the beginning of the section (p. 135).

$v(1) = 30 \quad v(2) = 12 \quad v(3) = 6 \quad v(1,2) = 36$

$v(1,3) = 36 \quad v(2,3) = 30 \quad v(1,2,3) = 90 \quad v(\emptyset) = 0$

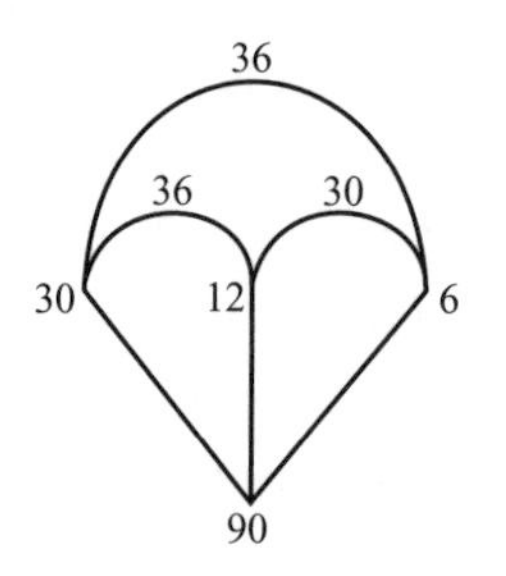

Suppose the coalition $\{2, 3\}$ claims its worth. We shall split the original game into a sum of two games:

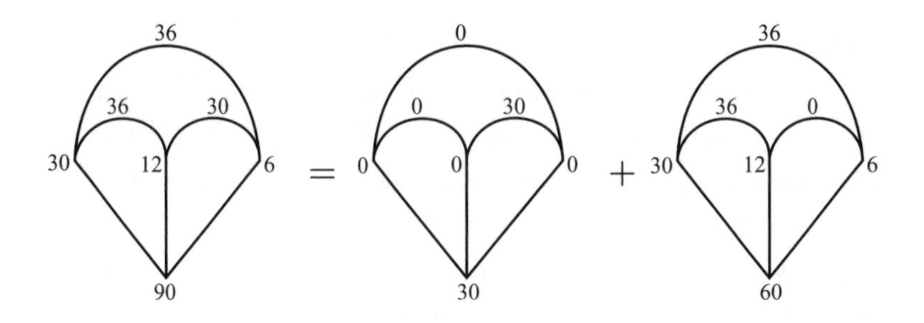

In the left-hand game, players 2 and 3 are symmetric and player 1 is a null player. Hence, 2 and 3 each get 15 and player 1 gets 0. Hence, the payoff $(0, 15, 15)$ is the Shapley value of the left-hand game, but it is also the division of payoffs according to the procedure for dissolving a partnership.

If we present it in a table, it will look like this:

		coalition worth							division		
claimant	game	{1}	{2}	{3}	{1,2}	{1,3}	{2,3}	{1,2,3}	1	2	3
{2,3}	v	30	12	6	36	36	30	90	0	0	0
	v_1	0	0	0	0	0	30	30	0	15	15
	u	30	12	6	36	36	0	60			

The game u is actually the difference between the two games v and v_1.

$$u = v - v_1 \Longleftrightarrow v = v_1 + u$$

As was said above, v_1 is a game in which players 2 and 3 are symmetric players and player 1 is a null player. Therefore, the Shapley value of the game is $(0, 15, 15)$, and that is precisely the division of payoffs. When coalition $\{2, 3\}$ claims its worth, they get the worth of their asset and divide it equally between themselves. In this way, every row in the table is obtained from the previous one. Hence, the procedure described above is actually the splitting of the original game into several carrier games in which all members of the carrier coalition are symmetric and the others are null players. Therefore, the final division of payoffs is the sum of the Shapley values of those games.

We have thus seen that if a joint enterprise is liquidated according to the procedure proposed above, the division of the proceeds of the sale of the enterprise among the partners is identical to the Shapley value of the game. That is, if we are obliged to find a fair division among partners of a joint enterprise that is about to be liquidated, then a division based on the Shapley value of the game is a reasonable proposal for such a division.

3.18 EXERCISES

1. A joint enterprise of three partners, 1, 2, and 3, consists of a travel agency, a car rental agency, a tour company, a souvenir shop, and a delivery service. The owner of the car rental agency, 1, forms a partnership with the owner of the travel agency, 2. Later the two purchase a souvenir shop, which they let their wives run. The owner of the tour company, 3, enters into a partnership with 2 and the two open a delivery service. Now 3 enters into the comprehensive partnership and the enterprise is the operation of all these businesses together.

Remark: We need to remember that $v(S)$ is the price that coalition S can achieve in the market for all the assets that belong either to it or to its subcoalitions. We shall describe the game in terms of the coalition function:

$$v(1) = 20 \quad v(2) = 30 \quad v(3) = 10 \quad v(1,2) = 40$$
$$v(1,3) = 0 \quad v(2,3) = 30 \quad v(1,2,3) = 60 \quad v(\emptyset) = 0$$

At a certain stage the joint enterprise starts losing money and the partners decide to liquidate it.

Calculate how much each of the partners will get from the sale of the whole enterprise, if they carry out the division according to the procedure for dissolving a partnership.

2. A joint enterprise consists of a perfume store, a clothing store, a jewelry store, and a leather goods store. The owner of the perfume

store, 1, partnered up with the owner of the clothing store, 2, and when their business prospered, they bought a jewelry store. Later they were joined by the owner of the leather goods store, 3. Afterwards, disagreements arose among the partners and they decided to dissolve the partnership.

Calculate how much each of the partners will get if the money is divided according to the procedure for dissolving a partnership and if the coalition function is:

$v(1) = 18 \quad v(2) = 36 \quad v(3) = 42 \qquad v(1,2) = 60$
$v(1,3) = 0 \quad v(2,3) = 0 \quad v(1,2,3) = 72 \quad v(\emptyset) = 0$

3. Find the Shapley value of the following game using the procedure for dissolving a partnership.

$N = \{1,2,3\}\ v(1) = 6 \quad v(2) = 6 \quad v(3) = 12 \quad v(\emptyset) = 0$
$v(1,2) = 18 \quad v(1,3) = 24 \quad v(2,3) = 0 \quad v(1,2,3) = 60$

4. Calculate the Shapley value of the following game using the procedure for dissolving a partnership.

$N = \{1,2,3,4\}$
$v(1) = 24 \qquad v(2) = 48 \qquad v(3) = 24 \qquad v(4) = 72$
$v(1,2) = 96 \qquad v(2,3) = 144 \qquad v(1,4) = 120 \quad v(1,3) = 0$
$v(2,4) = 0 \qquad v(3,4) = 0$
$v(1,3,4) = 168 \quad v(1,2,3) = 0 \qquad v(1,2,4) = 0$
$v(2,3,4) = 0 \qquad v(1,2,3,4) = 240 \quad v(\emptyset) = 0$

3.19 THE SHAPLEY VALUE AS THE AVERAGE OF PLAYERS' MARGINAL CONTRIBUTIONS

Consider the following game:

$N = \{1,2,3\}$
$v(1) = 6 \qquad v(2) = 12 \qquad v(3) = 18 \qquad v(\emptyset) = 0$
$v(1,2) = 30 \quad v(1,3) = 60 \quad v(2,3) = 90 \quad v(1,2,3) = 120$

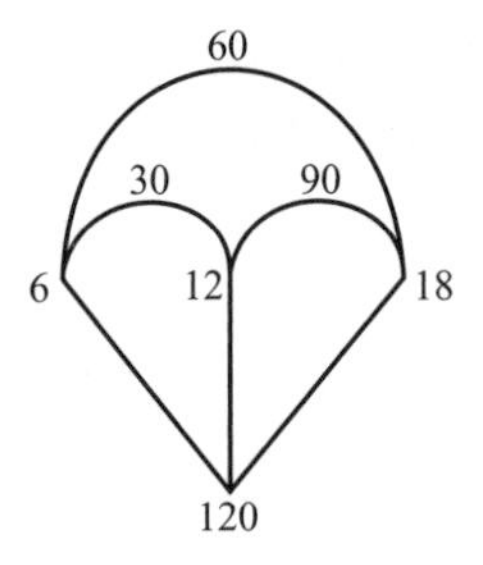

In Section 3.15 we found that the Shapley value of this game is $(22, 40, 58)$.

In this section we shall look at another way of finding the Shapley value.[9]

Let us start by imagining a procedure in which three players enter a room in any order. Each player that enters the room receives his marginal contribution to the coalition of players waiting for him in the room.

Example:

Suppose the players enter the room in the order $(2, 3, 1)$; that is, first player 2 enters the empty room. Before he entered there was an empty coalition in the room, whose worth was 0; after he enters, coalition $\{2\}$ is in the room, whose worth is 12, so 2 receives his marginal contribution, which is 12 units $(12 - 0)$.

Player 3 enters the room second. Player 2 is already in the room, so, together with player 3, coalition $\{2, 3\}$ is obtained, whose worth is 90. The worth of the coalition that was in the room before 3 arrived was 12, while the worth of the new coalition that is obtained upon 3's arrival is 90. Therefore, 3's marginal contribution is $90 - 12 = 78$, which is what player 3 gets.

Player 1 enters the room third and joins coalition $\{2, 3\}$, which is already in the room. When 1 arrives coalition $\{1, 2, 3\}$ is obtained, whose worth is 120; the worth of coalition $\{2, 3\}$, which was in the

[9] Shapley, L. S. 1953. "A value for n-person games," in Kuhn, H. and Tucker, A. W. (eds.), *Contributions to the theory of games II*, Annals of Mathematics Studies 28. Princeton: Princeton University Press, pp. 307–17

room beforehand was 90. Therefore, player 1's marginal contribution to the coalition is $120 - 90 = 30$ and he gets this contribution.

Summary:

order of entry into room			marginal contribution of player		
			1	2	3
2	3	1	30	12	78

The summary above refers only to a certain order, namely, $(2, 3, 1)$.

The procedure that we shall present below refers to a random order, where all possible orders have the same probability. The Shapley value is the *expectation* of the players in this procedure, that is, the average of the payoffs that they receive.

Let's reexamine the game described above. There are three players in the game and so there are 6 (= 3!) possible orders.

order of entry into room			marginal contribution		
			1	2	3
1	2	3	6	24	90
1	3	2	6	60	54
2	3	1	30	12	78
2	1	3	18	12	90
3	1	2	42	60	18
3	2	1	30	72	18
			132	240	348

The Shapley value of the game is:

$$\left(\frac{132}{6}, \frac{240}{6}, \frac{348}{6}\right) = (22, 40, 58).$$

Theorem:

The Shapley value of player i is the average of his marginal contributions that are paid out over all the possible orders.

The proof of the theorem goes beyond the scope of this book, but the theorem is of great importance, because it represents another aspect of the Shapley value.

Example: Two-Player Market Game

$$N = \{1, 2\} \quad v(1) = 2 \quad v(2) = 3 \quad v(1, 2) = 10$$

possible orders		marginal contribution	
		1	2
1	2	2	8
2	1	7	3
		9	11

The Shapley value of the game is $\left(\frac{9}{2}, \frac{11}{2}\right)$.

Example: Market with Two Buyers and Two Sellers

$$N = \{1, 2, 3, 4\} \quad v(1) = v(2) = 100 \quad v(3) = v(4) = 0$$
$$v(1, 2) = 200 \qquad v(3, 4) = 0$$
$$v(1, 3) = v(1, 4) = v(2, 3) = v(2, 4) = 150$$
$$v(1, 2, 3) = v(1, 2, 4) = 250$$
$$v(1, 3, 4) = v(2, 3, 4) = 150$$
$$v(1, 2, 3, 4) = 300$$

There are 4 players in the game and so there are 24 (4!) possible orders.

We shall list all the possible orders along with player 1's marginal contribution.

order	marg. cont.	order	marg. cont.
1234	100(= 100 – 0)	3124	150(= 150 – 0)
1243	100(= 100 – 0)	3142	150(= 150 – 0)
1324	100(= 100 – 0)	3214	100(= 250 – 150)
1342	100(= 100 – 0)	3241	150(= 300 – 150)
1423	100(= 100 – 0)	3421	150(= 300 – 150)
1432	100(= 100 – 0)	3412	150(= 150 – 0)
2134	100(= 200 – 100)	4123	150(= 150 – 0)
2143	100(= 200 – 100)	4132	150(= 150 – 0)
2314	100(= 250 – 150)	4213	100(= 250 – 150)
2341	150(= 300 – 150)	4231	150(= 300 – 150)
2413	100(= 250 – 150)	4321	150(= 300 – 150)
2431	150(= 300 – 150)	4312	150(= 150 – 0)

The total of player 1's marginal contributions over all the orders is 3000. The average of player 1's marginal contributions is $\frac{3000}{24} = 125$.

Players 1 and 2 are symmetric (verify it!), and so, by the symmetry axiom, the average of player 2's marginal contributions is 125 too.

The total payoff to players 1 and 2 is 250.

$v(1,2,3,4) = 300$ and so, by the efficiency axiom, 50 units of the payoff are left over to divide between players 3 and 4. Players 3 and 4 are symmetric too, so the amount of the payoff for each is 25.

To summarize, the Shapley value of the game is (125, 125, 25, 25).

3.20 EXERCISES

1. Calculate the Shapley value as the average of players' marginal contributions for the following games:

(1) Market game with two sellers and one buyer:

$$N = \{1,2,3\} \quad v(1) = 100 \quad v(2) = 100 \quad v(3) = 0 \quad v(\emptyset) = 0$$
$$v(1,2) = 200 \quad v(1,3) = 200 \quad v(2,3) = 200$$
$$v(1,2,3) = 300$$

(2)

$$N = \{1,2,3\} \quad v(1) = 0 \quad v(2) = v(3) = 6 \quad v(\emptyset) = 0$$
$$v(1,2) = 12 \quad v(1,3) = 6 \quad v(2,3) = 18$$
$$v(1,2,3) = 24$$

(3) Three-player weighted majority game:

$$[3; 2, 1, 1]$$

(4)

$$N = \{1,2,3\}$$
$$v(1) = 6 \quad v(2) = 12 \quad v(3) = 18 \quad v(\emptyset) = 0$$
$$v(1,2) = 12 \quad v(1,3) = 12 \quad v(2,3) = 24$$
$$v(1,2,3) = 48$$

2. Calculate the Shapley value of the following game:

$$N = \{1, 2\} \quad v(1) = a \quad v(2) = b \quad v(\emptyset) = 0$$
$$v(1, 2) = c$$

3. Calculate the Shapley value of the following games in two ways:

   (1) by dissolving a partnership;

   (2) by the average of players' marginal contributions.

   (i) $N = \{1, 2, 3\}$ $\quad v(1) = 10 \quad v(2) = 5 \quad v(3) = 0$
   $v(1, 2) = 18 \quad v(1, 3) = 10 \quad v(2, 3) = 6$
   $v(1, 2, 3) = 20 \quad v(\emptyset) = 0$

   (ii) $N = \{1, 2, 3\}$ $\quad v(1) = v(2) = 5 \quad v(3) = 0 \quad v(\emptyset) = 0$
   $v(1, 2) = 10 \quad v(1, 3) = v(2, 3) = 5$
   $v(1, 2, 3) = 10$

   (iii) $N = \{1, 2, 3\}$
   $v(1) = 6 \quad v(2) = v(3) = 12 \quad v(\emptyset) = 0$
   $v(1, 2) = v(1, 3) = 18 \quad v(2, 3) = 24$
   $v(1, 2, 3) = 30$

   (iv) $N = \{1, 2, 3, 4\}$
   $v(1) = v(2) = v(3) = 5 \quad v(4) = 0 \quad v(\emptyset) = 0$
   $v(1, 4) = v(2, 4) = v(3, 4) = 5$
   $v(1, 2) = 10 \quad v(1, 3) = v(2, 3) = 15$
   $v(1, 2, 3) = 20 \quad v(1, 2, 4) = 10 \quad v(1, 3, 4) = 15$
   $v(2, 3, 4) = 15$
   $v(1, 2, 3, 4) = 20$

   (v) $N = \{1, 2, 3, 4\}$
   $v(1) = v(2) = 6 \quad v(3) = v(4) = 12 \quad v(\emptyset) = 0$
   $v(1, 2) = 12 \quad v(3, 4) = 24$
   $v(1, 4) = v(2, 4) = 18 \quad v(1, 3) = v(2, 3) = 18$
   $v(1, 2, 3) = v(1, 2, 4) = 30$
   $v(1, 3, 4) = v(2, 3, 4) = 24$
   $v(1, 2, 3, 4) = 36$

3.21 THE SHAPLEY VALUE AS A PLAYER'S INDEX OF POWER IN WEIGHTED MAJORITY GAMES

In Section 3.7 we discussed weighted majority games whose general form is:

$$[q; w_1, w_2, ..., w_n]$$

$w_1, w_2, ..., w_n$ are the weights of the players. The weights are nonnegative numbers. q is the quota and we assume that it is greater than half the sum of the weights and less than or equal to the sum of the weights.

A coalition is called a *winning coalition* in such a game if the sum of the weights of its members is greater than q or equal to q. Otherwise, the coalition is called a *losing coalition*.

The coalition function of such a game is:

$$v(S) = \begin{cases} 1 & \text{if } S \text{ is winning} \\ 0 & \text{if } S \text{ is losing} \end{cases}$$

Exercise: Given an election for a city council, an interesting question arises: What is the political "strength" of each party? Any theory that attempts to measure such strength is called a *power index*.

One inclination is to say that a good index of power for a player i is the number of votes w_i he received in the election. This is not a satisfying definition. First, it is not influenced by the quota q. But more fundamentally, the game representation $(N; v)$ shows that the above suggestion is meaningless, because various weighted majority games of the form $[q; w_1, ..., w_n]$ lead to the *same game* $(N; v)$. We shall see shortly that the Shapley value of the representation $(N; v)$ is a reasonable definition of the index of power of that representation. This index is known as the *Shapley–Shubik power index*.[10]

First, we shall illustrate by way of example that a player's electoral power (i.e., his weight) is not a good index of his power in the game.

[10] Shapley, L. S. and Shubik, M. 1954. "A method for evaluation of the distribution of power in a committee system," *The American Political Science Review* 48: 787–92

Example:
Consider the game:

$$[8; 7, 1, 7]$$

In this example, the power of the second player exactly equals the power of the rest of the players, although he has only one representative while the others each have 7 representatives. To pass a decision, a two-player coalition is necessary. Indeed, the coalition function of the game is:

$$v(S) = \begin{cases} 0 & \text{if } S \text{ is empty or has 1 player} \\ 1 & \text{if } S \text{ has 2 or 3 players} \end{cases}$$

We see, then, that no preference exists for one of the players, 1 or 2 or 3, and all the players are symmetric (verify it!).

By the efficiency and symmetry axioms the Shapley value of the game is $(\frac{1}{3}, \frac{1}{3}, \frac{1}{3})$.

The Shapley value shows that the players' power in the game is equal, although their electoral representation is different.

The question is whether the Shapley value constitutes a good index of the players' power in other weighted majority games.

In this section we shall see that, in a certain sense, it does. We shall ask what we mean by the Shapley value in weighted majority games. As we know, the Shapley value of every player is the average of his marginal contributions over all possible orders.

We shall start with the game $[10; 5, 8, 2, 3]$ and calculate the marginal contributions of all players when the order is $1, 2, 3, 4$. Player 1's marginal contribution in this order is 0, because $\{1\}$ is a losing coalition. Player 2's marginal contribution in this order is 1, because $\{1, 2\}$ is a winning coalition. Player 3's marginal contribution in this order is 0, because in joining $\{1, 2\}$ player 3 does not increase its worth. Player 4's marginal contribution in this order is also 0, because in joining $\{1, 2, 3\}$ player 4 does not increase its worth. (Explain!)

We thus see that, in this order, player 2's marginal contribution is 1 and the rest of the players' marginal contribution is 0.

Player 2 will be called a *pivotal player in this order.*

Let's try another order, say 3, 4, 1, 2:
Player 3's marginal contribution is 0.
Player 4's marginal contribution is 0.
Player 1's marginal contribution is 1.
Player 2's marginal contribution is 0.

We see that, in this order, player 1 is the only player whose contribution is 1. The other players' marginal contribution is 0. Player 1 is a *pivotal player in this order.*

In any weighted majority game there is exactly one pivotal player in every order. This is the player who in joining a coalition turns it from a losing coalition to a winning coalition. Before he joined the coalition, the players who were before him in order did not contribute anything to the coalition worth; after he joins the coalition, the players who come after him in order do not contribute anything to the coalition worth.

To summarize: a player is called a pivotal player in a certain order, if his marginal contribution in that order is 1.

Theorem:
In a weighted majority game, there is exactly one pivotal player in every order.
Proof: First we shall see that there is a player who by joining the players who were before him turns a losing coalition to a winning coalition. According to the properties of the coalition function in a weighted majority game, an empty coalition is a losing coalition. Then the players join it one after the other, in a given order. If, during this procedure, not one of the players turns the coalition of the players who were before him to a winning coalition, then the set of all players is a losing coalition, which contradicts a property of the coalition function, according to which the set of all players is a winning coalition. The contradiction proves the existence of a player who by joining the coalition of the players who were before him turns it from a losing

coalition to a winning coalition. The marginal contribution of this player is 1. The marginal contribution of all the players who were before him is 0 and the marginal contribution of all the players who were after him is 0. Hence this player is the only pivotal player.

As we know, a player's Shapley value is the average of his marginal contributions over all the possible orders. We can therefore summarize the above in the following theorem:

Theorem:

In a weighted majority game of n players, a player's Shapley value is:

$$\text{Frequency with which player is pivotal over all possible orders} = \frac{\text{number of times player is pivotal}}{\text{number of possible orders } (n!)}.$$

Let us calculate the Shapley–Shubik power index of player 1 in the game $[10; 5, 8, 2, 3]$:

- When he is first, his marginal contribution is 0. (Explain!) There are 6 such orders.
- When he is second, his marginal contribution is 1 only if player 2 is before him. There are two such orders $(2, 1, 3, 4; 2, 1, 4, 3)$.
- When he is third, his marginal contribution is 1 only if player 2 is after him. (Explain!) There are two such orders.
- When he is fourth, his marginal contribution is 0. There are six such orders.

Hence, the Shapley–Shubik index of player 1 is $\frac{4}{24} = \frac{1}{6}$.

Exercise: Calculate the Shapley–Shubik index of all players in the game above.

Setting aside the Shapley–Shubik index for a moment, let us consider the following practical situation modeled on the game $[10; 5, 8, 2, 3]$.

There is a city council in which four parties are represented by 5, 8, 2, and 3 representatives, respectively. How shall we estimate the power of, say, the first party?

One way to estimate a party's power is to check how frequently the party's vote is needed to pass a law.

Suppose that a certain law is proposed to the city council and support for it is split as follows:

Party 2	Party 3	Party 1	Party 4
strongly in favor	in favor	undecided	opposed

In this case the vote of party 1 is not needed, because the support of parties 2 and 3 is sufficient for a decision in favor of the law.

Note that under this arrangement party 1 is not a pivotal player.

Suppose that the amount of support for another law is split as follows:

Party 2	Party 1	Party 3	Party 4
in favor	undecided	opposed	strongly opposed

In this case party 1 needs to be persuaded to support the law, because its vote is needed to establish a majority in favor of the law. Party 2 will definitely try to persuade party 1 to support the law.

Note that under this arrangement party 1 is a pivotal player.

Since we do not know in advance what laws will be proposed, we shall introduce the following assumption, which may be reasonable in many cases.

Assumption:[11]
Several different laws will be proposed, such that every possible order, which represents an amount of support, has an equal chance of occurrence.

We shall think of the possible orders as if they were a deck of cards, such that every time we shuffle the deck every order has an equal chance of occurrence.

It follows from this assumption that how frequently a party's vote is needed to establish a majority in favor of any law exactly equals how frequently the party constitutes a pivotal player over all possible orders.

[11] This assumption is unreasonable when, for example, the proposed laws are of a political character and the players are parties with well-defined political platforms.

It is reasonable to measure the party's strength by this frequency and it follows that the Shapley–Shubik index in a weighted majority game is a reasonable index of the parties' power.

3.22 EXERCISES

Calculate the Shapley–Shubik index in the following games:

1. $[10; 7, 5, 4, 3]$
2. $[12; 4, 4, 9, 5]$
3. $[17; 7, 8, 9, 9]$
4. $[7; 4, 2, 2, 2, 2]$
5. $[5; 3, 3, 1, 1, 1]$
6. $[9; 4, 4, 2, 2, 2, 2]$
7. $[6; 3, 1, 1, 1, 1, 1, 1, 1]$

3.23 THE SHAPLEY–SHUBIK INDEX AS AN INDEX FOR THE ANALYSIS OF PARLIAMENTARY PHENOMENA

Take, for example, a parliament that consists of five parties. One party has 40 representatives and each of the other four parties has 20 representatives. The game is $[61; 40, 20, 20, 20, 20]$.

Calculate the Shapley–Shubik index of player 1:

- When he is first, he is not a pivotal player so his marginal contribution is 0.
- When he is second, he is not a pivotal player so his marginal contribution is 0.
- When he is third, he is a pivotal player so his marginal contribution is 1. There are 4! such orders. Indeed, when player 1 is third, there are four players left. One of them can be placed in one of four positions. For each such placement, another can be placed in three positions. For each such placement, the third has only two options and after he is placed, the fourth has only a single option. Altogether, we have $4 \cdot 3 \cdot 2 \cdot 1 = 4!$ possible arrangements, when player 1 is third.

- When he is fourth, he is a pivotal player so his marginal contribution is 1. There are 4! such orders.
- When he is fifth, he is not a pivotal player so his marginal contribution is 0.

Therefore, the Shapley-Shubik index of player 1 is $\frac{4!+4!}{5!} = \frac{48}{120} = \frac{2}{5}$.

Players $2, 3, 4$, and 5 are symmetric, so their Shapley-Shubik index is equal. By the efficiency axiom, $v(N) = 1$ needs to be divided among all the players. Hence, the Shapley-Shubik index of the game is:

$$\left(\frac{2}{5}, \frac{3}{20}, \frac{3}{20}, \frac{3}{20}, \frac{3}{20}\right).$$

We shall examine several other cases in which there is one party with 40 representatives and the number of other, smaller parties grows each time.

1. $[61; 40, 10, 10, 10, 10, 10, 10, 10, 10]$.

The Shapley-Shubik index of the game is:
$\left(\frac{4}{9}, \frac{5}{72}, \frac{5}{72}, \frac{5}{72}, \frac{5}{72}, \frac{5}{72}, \frac{5}{72}, \frac{5}{72}, \frac{5}{72}\right)$.

2. $[61; 40, 8, 8, 8, 8, 8, 8, 8, 8, 8, 8]$.

The Shapley-Shubik index of the game is:
$\left(\frac{5}{11}, \frac{3}{55}, \frac{3}{55}, \frac{3}{55}, \frac{3}{55}, \frac{3}{55}, \frac{3}{55}, \frac{3}{55}, \frac{3}{55}, \frac{3}{55}, \frac{3}{55}\right)$.

3. $q = 61 \qquad w_1 = 40 \qquad w_2 = w_3 = \ldots = w_{17} = 5$.

The Shapley-Shubik index is $\left(\frac{8}{17}, \frac{9}{272} \cdots, \frac{9}{272}\right)$.

4. $q = 61 \qquad w_1 = 40 \qquad w_2 = w_3 = \ldots = w_{21} = 4$.

The Shapley-Shubik index is $\left(\frac{10}{21}, \frac{11}{420} \cdots, \frac{11}{420}\right)$.

5. $q = 61 \qquad w_1 = 40 \qquad w_2 = w_3 = \ldots = w_{41} = 2$.

The Shapley–Shubik index is $\left(\frac{20}{41}, \frac{21}{1640} \cdots, \frac{21}{1640}\right)$.

6. $q = 61 \qquad w_1 = 40 \qquad w_2 = w_3 = \ldots = w_{81} = 1$.

The Shapley–Shubik index is $\left(\frac{40}{81}, \frac{41}{6480} \cdots, \frac{41}{6480}\right)$.

In each case there is one big party which has $\frac{1}{3}$ of the representatives. We see that the greater the number of small parties, the greater the strength of the big party, which approaches $\frac{1}{2}$.[12]

$$\frac{1}{3} < 0.4 < 0.44 < 0.45 < 0.47 < 0.476 < 0.487 < 0.49$$

At the same time, the small parties have little power compared with their electoral power. In the first game, for example, a small party has 8.3% of the representatives, while its power according to the Shapley–Shubik index is 6.9%.

In the fourth game, each small party has 3.3% of the representatives and its power is 2.6%.

This is a general phenomenon: if there is one big party and a large number of small parties, then usually the big party has more power than one would expect on the basis of the number of its representatives alone. The greater the disunity of the small parties, the greater the power of the big party.

If the number of representatives in the big party is 45 and the number of small parties increases, we see that the power of the big party approaches 60%.

We shall calculate the Shapley–Shubik index in another situation and see that this time we get a less expected result.

Consider the possibility that an alternative to the big party is established, such that there will be two big parties in parliament, and assume that each of the parties has 40 representatives, while the rest of the representatives are divided among 5 parties, such that each one has 8 representatives.

The Shapley–Shubik index of the game $[61; 40, 40, 8, 8, 8, 8, 8]$ is:

$$\left(\frac{2}{7}, \frac{2}{7}, \frac{3}{35}, \frac{3}{35}, \frac{3}{35}, \frac{3}{35}, \frac{3}{35}\right).$$

It turns out that while each big party has $\frac{40}{120} \sim 33\%$ of the votes, its power is only $\frac{2}{7} \sim 29\%$. Similarly, while each small party has

[12] Milnor, J. and Shapley, L. S. 1961. "Values of large games II: oceanic games," The Rand Corporation, Memorandum RM-2649

$\frac{8}{120} \sim 7\%$ of the votes, it controls $\frac{3}{35} \sim 9\%$ of the power. Put differently, a representative of a big party controls $\frac{1}{40} \cdot 0.29\% \sim 0.7\%$, whereas a representative of a small party controls $\frac{1}{8} \cdot 9\% \sim 1.125\%$ – almost twice as much!

In this case, the increase in the small parties' power is precisely the result of their disunity. In other words, if the small parties were united, then we would have 3 symmetric players, so the small parties would hold 33% of the power after uniting, compared to about 43% of the power in a situation of disunity. Moreover, the greater the disunity of the small parties, the greater their power: if they are very small, the power of each of the big parties decreases to 25%, while the small parties hold about 50% of the power. In this case we can say, therefore, that in disunity there is strength; in unity, weakness.

3.24 EXERCISES

Calculate the Shapley–Shubik index of the following games:

1. $q = 61 \quad w_1 = 45 \quad w_2 = w_3 = \ldots \quad w_{16} = 5$
2. $q = 61 \quad w_1 = 45 \quad w_2 = w_3 = \ldots \quad w_{26} = 3$
3. $q = 61 \quad w_1 = 45 \quad w_2 = w_3 = \ldots \quad w_{76} = 1$
4. $q = 61 \quad w_1 = w_2 = 40 \quad w_3 = w_4 = \ldots \quad w_{12} = 4$
5. $q = 61 \quad w_1 = w_2 = 40 \quad w_3 = w_4 = \ldots \quad w_{22} = 2$
6. $q = 61 \quad w_1 = w_2 = 40 \quad w_3 = w_5 = \ldots \quad w_{42} = 1$

3.25 THE SECURITY COUNCIL

The Security Council has fifteen members: the five Big Powers (United States, Russia, France, England, and China) which are permanent members and have veto power, and ten small states which are rotating members elected by the General Assembly for two-year terms. Each Council member has one vote. Decisions on all matters require at least nine votes; decisions on substantive matters require in addition the concurring votes of all five permanent members. No important decision can be made without the agreement of the five permanent members. A negative vote by a permanent member constitutes a veto. An abstention is not considered a veto in most cases.

Using the Shapley–Shubik index we can see how veto power is implemented; that is, we can see what the power of a permanent member of the Security Council is, and compare it to the power of a rotating member.

Consider the Security Council as a weighted majority game whose five permanent members are symmetric and whose ten small states are symmetric. It follows that in calculating the Shapley–Shubik index we have to consider only two different values. (Explain!)

We check the aggregate power of all the small states, namely, the number of orders in which a small state is a pivotal player, divided by the number of all orders.

An order in which a small state is a pivotal player occurs if and only if seven small states come last in the order. (Explain, using the following figure.)

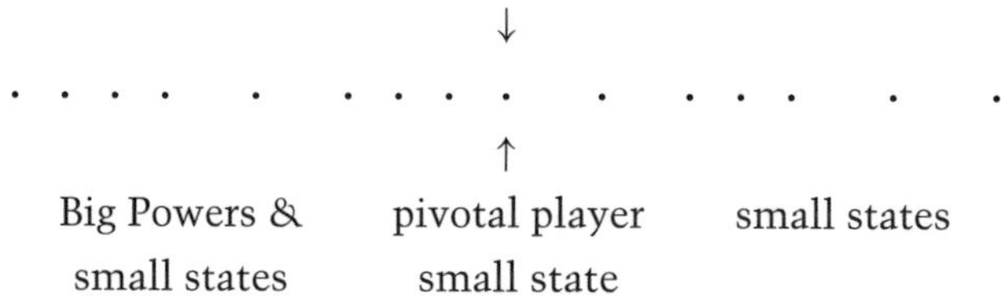

↓

. . . .

↑

Big Powers & small states	pivotal player small state	small states

To obtain all such orders one has to choose seven small states out of ten and then calculate the number of cases with this combination.

The number of combinations of seven small states from ten is given by the formula

$$\binom{10}{7} = \frac{10!}{(10-7)! \cdot 7!} = \frac{10!}{3! \cdot 7!}$$

(Here $k! = 1 \times 2 \times 3 \times \ldots \times k$.)

For each such order we can permute, in any way we wish, the order of the last seven states, as well as the order of the first eight states. Thus, the number of all $\frac{10!}{3!\cdot 7!}$ orders has to be multiplied by $7!\cdot 8!$. Consequently, the number of orders in which seven small states come last is

$$\frac{10!\cdot 7!\cdot 8!}{3!\cdot 7!} = \frac{10!\cdot 8!}{3!}$$

Thus, the aggregate power of the small states is

$$\frac{10!\cdot 8!}{3!\cdot 15!} \approx 0.0186 = 1.86\%$$

(since 15! is the number of all orders). Therefore, the index of power of a single small state is

$$\frac{0.0186}{10} = 0.00186 = 0.186\%$$

and the aggregate of power of permanent members is equal to

$$100\% - 1.86\% = 98.14\%.$$

Consequently, the index of power of a single permanent member is equal to

$$\frac{98.14}{5} \approx 19.63\%.$$

3.26 EXERCISES

1. What would be the power of all the small states if a simple majority were customary in the Security Council, that is, if at least eight votes including the concurring votes of the five permanent members were necessary to pass a resolution?

2. Until 1965, the Security Council consisted of eleven members: five permanent members – the five Big Powers – and six rotating members – the small states. In this makeup of the Security Council a majority of seven was required to pass a resolution. What was the aggregate

power of the small states in this situation? What was the power of each permanent member?

3. What is the aggregate power of the small states in a Security Council of eleven members if a simple majority is required to pass a resolution, that is, if at least six votes are sufficient including the concurring votes of the five permanent members?

3.27 COST GAMES

The focus of this chapter has been the mathematical model of the cooperative game $(N; v)$, where N is the set of players and v is the coalition function that assigns to each coalition S a real number $v(S)$ which is the total payoff that coalition S will get if its members decide to form it.

Similarly, we can construct a mathematical model for a cooperative game $(N; c)$, where N is the set of players and c is the coalition function that assigns to each coalition S a real number $c(S)$ which is, instead, the total *cost* that coalition S will pay if its members decide to form it.

Example:

A cable TV station wants to connect customers into a network. Network connections can be described in a graph called a "tree."[13] The cost of establishing each arc in the network appears in the figure above the arc. The players appear in the vertices of the tree. (A vertex with no players is called a "junction.")

Consider the game $(N; c)$ when $N = \{1, 2, 3\}$. The total cost that coalition S will incur if only its members connect to the station is $c(S)$.

[13] A *connected graph* is a figure consisting of vertices and arcs such that a path is traced in arcs from one vertex to another. A connected graph is called a *tree* if it contains no cycles, i.e., if a *unique* path is traced in arcs from one vertex to another such that no arc is traced more than once.

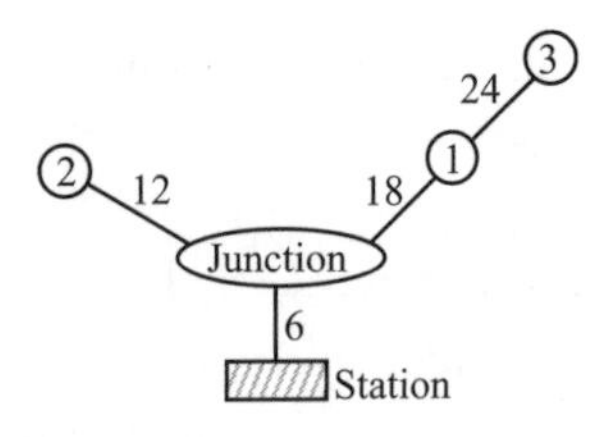

We see from the tree that the game has the following coalition function:

$$N = \{1, 2, 3\} \quad c(1) = 24 \quad c(2) = 18 \quad c(3) = 48 \quad c(\emptyset) = 0$$
$$c(1, 2) = 36 \quad c(1, 3) = 48 \quad c(2, 3) = 60$$
$$c(1, 2, 3) = 60$$

If all three connect to the network, that is, if coalition $N = \{1, 2, 3\}$ forms, then it can be asked how the total cost should be distributed among them. One answer is the Shapley value of the game.

We shall calculate the Shapley value as the average of the players' marginal contributions.

	1	2	3
1 2 3	24	12	24
1 3 2	24	12	24
2 1 3	18	18	24
2 3 1	0	18	42
3 1 2	0	12	48
3 2 1	0	12	48
	66	84	210

The Shapley value of the game is $(11, 14, 35)$.

Interestingly, these numbers can be obtained by a different rule: *Divide the cost of each arc equally among the players who use the arc*. Indeed, implementing this rule we get:

	1	2	3
Everyone uses the first segment:	2	2	2
Only player 2 uses the segment costing 12:	0	12	0
Only players 1, 3 use the segment costing 18:	9	0	9
Only player 3 uses the segment costing 24:	0	0	24
Total:	11	14	35

This rule is valid for every tree game. It enables us to compute the Shapley value for tree games with many players easily. The proof of this rule falls outside the scope of this book.

Example:
A cable TV station wants to connect nine customers to its network. Network connections are described in the following tree graph and as in the previous example the cost of establishing each arc in the network appears above the arcs and the players appear in the vertices of the tree. In the game $N = \{1, 2, ..., 9\}$, as in the previous example, $c(S)$ is the total cost that coalition S incurs if only its members connect to the station.

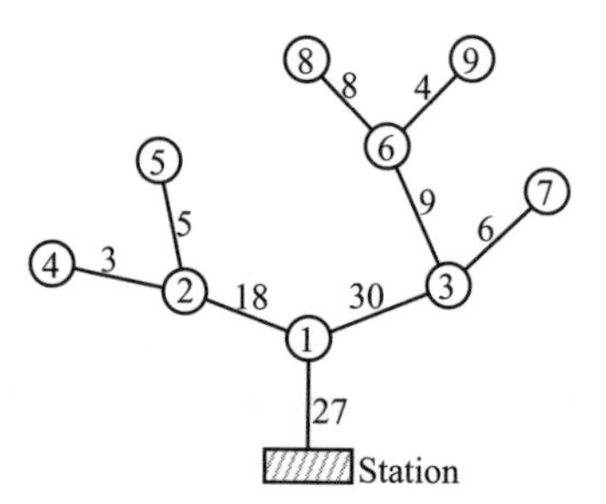

Even writing the coalition function involves $2^9 = 512$ worths, and so the number of all possible orders needed to compute the Shapley value is virtually incalculable. Using the above rule is much easier.

	1	2	3	4	5	6	7	8	9
All (players) use the first segment:	3	3	3	3	3	3	3	3	3
Only 2, 4, 5 use the segment costing 18:		6		6	6				
Only 4 uses the segment costing 3:				3					
Only 5 uses the segment costing 5:					5				
Only 3, 6, 7, 8, 9 use the segment costing 30:			6			6	6	6	6
Only 7 uses the segment costing 6:							6		
Only 6, 8, 9 use the segment costing 9:						3		3	3
Only 8 uses the segment costing 8:								8	
Only 9 uses the segment costing 4:									4
	3	9	9	12	14	12	15	20	16

The Shapley value of the game is $(3, 9, 9, 12, 14, 12, 15, 20, 16)$.

Discussion Question: Review all the axioms that establish the Shapley value of a game and determine whether they make sense in reference to a cost game.

3.28 EXERCISES

1. A cable TV station wants to connect three new customers to its network. The connections to the network are described in the following tree graph:

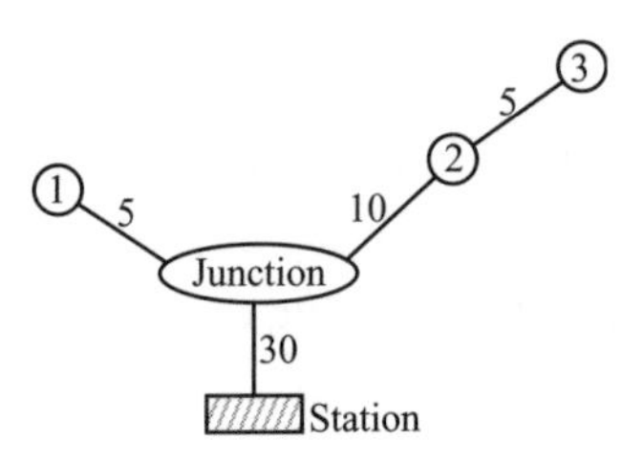

Calculate the distribution of the costs among the three players according to the Shapley value.

2. A cable TV station wants to connect three new customers to its network. The connections to the network are described in the following tree graph:

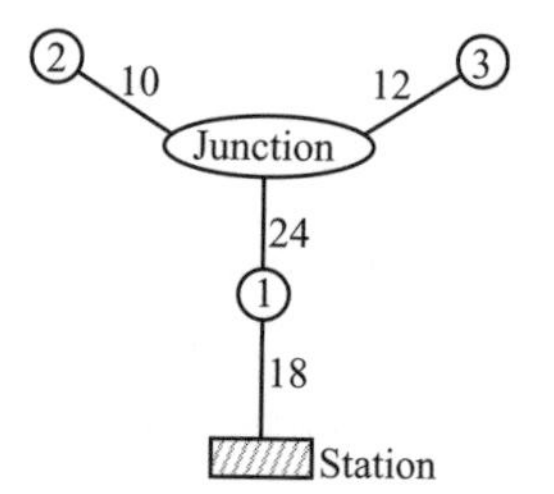

Calculate the distribution of the costs among the three players if all three decide to connect to the network.

3. A cable TV station wants to connect six new customers to its network. The connections to the network are described in the following tree graph:

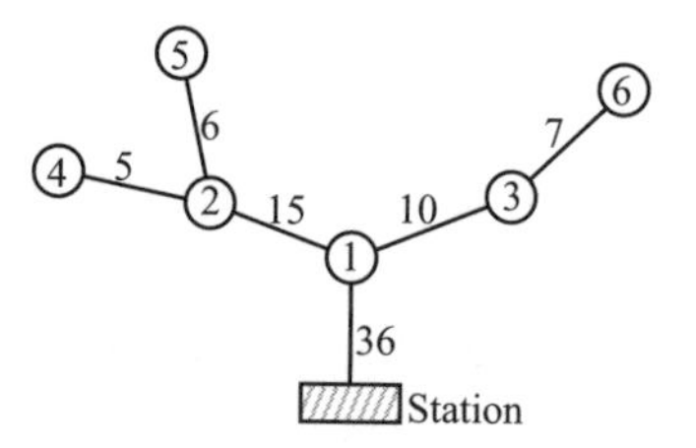

Calculate the distribution of the costs among the six players if all six decide to connect to the network.

4. Residents of towns near a city are responsible for maintenance of the roads linking them to the city. The following tree describes the data. Each town is represented by a vertex and the number of residents appears in the vertex. The monthly cost of maintenance for each section of road appears above the arc of the tree.

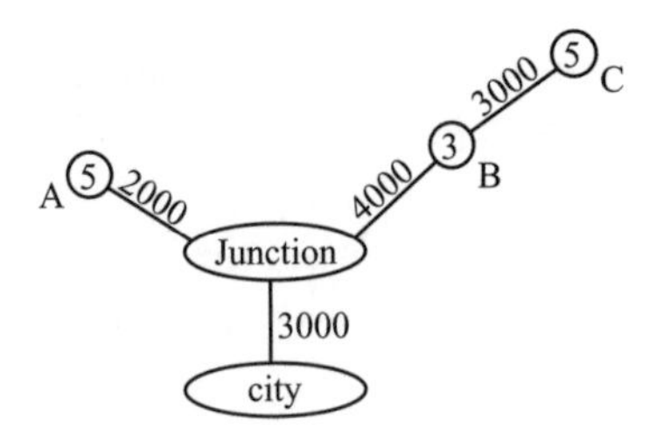

Calculate the distribution of the costs among all residents, using the Shapley value.

5. A special case of the tree game is the *airport game.* A runway is divided into three sections. The first section is for the use of light aircraft and the cost of establishing it is \$120,000. Medium aircraft need a longer runway and the cost of establishing the extra section is \$60,000. Heavy aircraft need an even larger runway and the cost of establishing the additional section is \$20,000. This is presented in the following figure:

Terminal
120,000
60,000
20,000
L_1, L_2
M_1
M_2
M_3
H

Suppose there are two light aircraft L_1, L_2, three medium aircraft M_1, M_2, M_3, and one heavy aircraft H at this airport. Every landing is considered a "player." The coalition function $c(S)$ is the cost of the runway needed to serve all players in S.

(1) Calculate: $c(L_1) =$ $\quad c(L_1, L_2) =$ $\quad c(L_1, H) =$ $\quad c(M_1, M_2, H) =$
(2) Calculate the distribution of costs among all flights according to the Shapley value.

3.29 REVIEW EXERCISES

1. Given the game $[61; 40, 40, 30, 10]$:

   (1) Write the game as a coalition function.
   (2) Are there any symmetric players? If so, which ones?

(3) Is there a null player? If so, which one?
(4) Calculate the Shapley–Shubik power index for the game.

2. Given the game $(N; v)$:

$N = \{1, 2, 3\}$

$$v(1) = 6 \quad v(2) = 6 \quad v(3) = 12 \quad v(\emptyset) = 0$$
$$v(1,2) = 18 \quad v(2,3) = 12 \quad v(1,3) = 24 \quad v(1,2,3) = 42$$

(1) Split the game into two games, such that in one of the games players 1 and 2 are symmetric and player 3 is a null player.
(2) Is there a null player and/or symmetric players in the second game too?

3. Calculate the Shapley–Shubik index of the following game: $[4; 3, 1, 1, 1]$.

4. Calculate the Shapley value of the following game:

$N = \{1, 2, 3\}$

$$v(1) = v(2) = 50 \quad v(3) = 0 \quad v(\emptyset) = 0$$
$$v(1,2) = v(1,3) = v(2,3) = 100$$
$$v(1,2,3) = 150$$

5. Find the Shapley value by the procedure for dissolving a partnership.

$N = \{1, 2, 3\}$

$$v(1) = v(2) = 12 \quad v(3) = 18 \quad v(\emptyset) = 0$$
$$v(1,2) = 18 \quad v(1,3) = 24 \quad v(2,3) = 30$$
$$v(1,2,3) = 60$$

6. Calculate the Shapley–Shubik index as the average of players' marginal contributions in the following game: $[4; 3, 2, 1, 1]$.

7. Calculate the Shapley value of the following game:

$N = \{1, 2, 3\}$

$$v(1) = 6 \quad v(2) = 12 \quad v(3) = 18 \quad v(\emptyset) = 0$$
$$v(1,2) = 18 \quad v(1,3) = 24 \quad v(2,3) = 30$$
$$v(1,2,3) = 42$$

8. Calculate the Shapley–Shubik index of the following game: $[8; 3, 3, 2, 2, 2, 2]$.

4 Analysis of a Bankruptcy Problem from the Talmud

4.1 INTRODUCTION

Many times one encounters a bankruptcy situation where there are claims against a given estate and the sum of the claims against the estate exceeds its worth. In such situations one would like to know what would be a "fair" way of dividing the estate among the claimants.

Unfortunately, there is no clear-cut answer to this question. What seems fair in one case may seem less so in another. In this chapter we shall encounter several solutions, each shedding light on the "real world" and each applicable under certain circumstances.

We start with a curious method of division that has its origin in the Talmud,[1] which represents still another fair division. It involves a man who married three women and promised them in their marriage contract the sums of 100, 200, and 300 units of money to be given to them upon his death. The man died but his estate amounted to less than 600 units. The Mishna, attributed to Rabbi Nathan (tractate Ketubot 93a), treats the cases in which the estate was worth 100, 200, and 300 units of money. The recommendation in the Mishna is given in the following table.

Claims \ Estate	100	200	300
100	33⅓	50	50
200	33⅓	75	100
300	33⅓	75	150

This recommendation of Rabbi Nathan seems strange. Why equal division if the estate is small? Why proportional division if the estate is worth 300 units? Most strangely, how did Rabbi Nathan reach

[1] An ancient document that forms the basis for Jewish religious, criminal, and civil law. It consists of the *Mishna*, which is its core, and the *Gemara*, which discusses the Mishna and expands on it. The Mishna was put into definitive form about 1800 years ago and the Gemara was sealed about 200 years later.

the division for the case in which the estate is worth 200? Above all, what should the rule be if the worth of the estate were different and if there were more widows?

Indeed, for many years this passage was not understood, and different rules of division were adopted by different rabbinic scholars. Some thought that this division reflected special circumstances whose description was neglected. Another thought that there was a spelling mistake. The wording of the Talmud itself suggests that this recommendation was not adopted, and that a different law was applied. One important rabbinic scholar, Hai Gaon, expressed the opinion that there might be some relation between this rule and the rule for dividing a garment between two claimants (see Section 4.2). However, Rabbi Hai Gaon did not explain the relation, and eventually retracted his opinion.

Despite myriad discussions among various scholars, no solid explanation was found until quite recently. Two game theorists, R. J. Aumann and M. Maschler, examined the rule. They decided to translate the three bankruptcy problems into game models and see if known solution concepts would yield the results stated in the Mishna. To their surprise, they found that one solution concept, called the *nucleolus*, gave precisely the numbers of the above table. It seemed that, finally, an explanation of Rabbi Nathan's recommendation had been found. There was only one "minor" problem: the nucleolus was invented by D. Schmeidler[2] in 1969. It was absolutely inconceivable that Rabbi Nathan knew what the nucleolus was.[3] There had to be another explanation for the numbers in the table. A hint was found in a paper by the game theorist A. I. Sobolev, who provided a system of axioms that characterize the nucleolus.[4] One of these axioms, called *consistency*, was the right clue.

[2] Schmeidler, D. 1969. "The nucleolus of a characteristic function game," *SIAM Journal of Applied Mathematics* 17: 1163–70

[3] A description of the nucleolus is beyond the scope of this book.

[4] Sobolev, A. I. 1975. "The characterization of optimality principles in cooperative games by functional equations," in Vorobiev, N. N. (ed.), *Matematicheskie Metody v Socialnix Naukax* 6. Academy of Sciences of the Lithuanian S. S. R., Vilnius, pp. 94–151

In this chapter we explain the concept of consistency and show how it yields a reasonable explanation of Rabbi Nathan's table. Moreover, it shows clearly how similar problems with more creditors and various claims can be resolved.[5]

To understand this explanation we first have to understand another, simpler Mishna rule involving a contested garment.

4.2 THE CONTESTED GARMENT

The following Mishna appears in the Talmud (tractate Bava Metzia 2a): "Two hold a garment; both claim it all. Then the one is awarded half, the other half. Two hold a garment; one claims it all, the other claims half. Then the one is awarded 3/4, the other 1/4."

We shall now discuss the claims and the decision of this Mishna. In the first case, both sides claim the whole garment and the decision establishes that in this case each claimant gets half the length of the garment.

The second case is of much greater interest to us. The one claims the whole garment and the other claims half. In this case the decision establishes that the claimant to the whole garment receives 3/4 of it and the claimant to half the garment receives 1/4.

How was this division reached? Rabbi Shlomo Yitzhaki (Rashi) interprets the decision as follows. The claimant to half the garment "concedes ... that half belongs to the other, so that the dispute revolves solely around the other half. Consequently, ... each of them receives half the disputed amount." Thus it is decided that the division shall be 3/4 and 1/4.

In this section we shall generalize the problem to other cases.

Example 1

The garment is worth 100 units of money.
One claims that his share of the garment is 50 units.
The other claims that his share of the garment is 80 units.
How should they divide it?

[5] Aumann, R. J. and Maschler, M. 1985. "Game-theoretic analysis of a bankruptcy problem from the Talmud," *Journal of Economic Theory* 36: 195–213

Solution:

The claimant to 50 units of money declares in effect that he has no claim to the second 50 units, and, as far as he is concerned, the other claimant can have them. The claimant to 80 units declares that he has no claim to the remaining 20 units, and, as far as he is concerned, the first claimant can have them. Thus uncontested, 70 of the 100 units are divided. The division therefore revolves around the remaining 30 units of money, which are to be divided equally between the two. The description of the division is as follows.

Value of garment	100	
The two claims	80	50
Uncontested division	50	20
Equal division of remainder	15	15
	65	35

The claimant to 50 units gets 35 and the claimant to 80 units gets 65.

Example 2

A man has two creditors; one's claim is 300, the other's, 90. The man's estate is worth 120 units. This is a bankruptcy problem. We shall solve it according to the "contested-garment" principle.[6]

Estate	120	
Claims	90	300
Uncontested division	0	30
Equal division of remainder	45	45
	45	75

[6] We are taking the position that any claim greater than the estate should be truncated to the size of the estate since there is nothing more to divide.

Answer:

The claimant to 300 units gets 75 and the claimant to 90 units gets 45.

A new element appears in Example 2. One of the debts exceeds the total amount available for distribution. It is worth noting that the creditors address their claims to the debtor and not to each other. The claimant to 90 units has no claim on the remaining 30 units. As far as he is concerned, those 30 units can be paid to the other creditor. On the other hand, the claimant to 300 units in effect claims the entire estate. Unfortunately for him, he cannot claim more than that amount, because there is no additional property. As far as he is concerned, there is no money left that he does not claim, and so, from his standpoint, there is nothing left for the other claimant, which explains the 0 that appears in the column of the claimant to 90 units.

Mathematical generalization:

The estate is E.

The creditors claim d_1 and d_2.

$d_1 + d_2 > E$; otherwise there is nothing to prevent full repayment of the debt.

Division of the estate is as follows.

Estate	E	
Claims	d_1	d_2
Uncontested division	$(E - d_2)_+$	$(E - d_1)_+$
Equal division of remainder	$\frac{E-(E-d_1)_+-(E-d_2)_+}{2}$	$\frac{E-(E-d_1)_+-(E-d_2)_+}{2}$
	$\frac{E-(E-d_1)_++(E-d_2)_+}{2}$	$\frac{E+(E-d_1)_+-(E-d_2)_+}{2}$

Explanation: The plus sign (+) in the expression $(E-d_1)_+$ or $(E-d_2)_+$ means that the expression has a value of zero if $E-d_1 < 0$ or $E-d_2 < 0$.

4.3 EXERCISES

1. A garment is worth 150 units of money. One claims 75 units, the other claims 100 units. How should they divide the garment, according to the contested-garment principle?

2. A garment is worth 200 units. One claims 120 units, the other claims 180 units. How should they divide the garment, according to the contested-garment principle?

3. A man goes bankrupt and his entire estate at the time of bankruptcy is worth 200 units. The man has two creditors; one's claim is 300 units, the other's, 200 units. How should they divide the estate between them, according to the contested-garment principle?

4. A man goes bankrupt and his entire estate at the time of bankruptcy is worth 300 units. The man has two creditors; one's claim is 250 units, the other's, 130 units. How should they divide the estate between them, according to the contested-garment principle?

5. A man dies, leaving an estate worth 500 units. The deceased has two creditors; one's claim is 400 units, the other's, 300 units. The division of the estate between them is as follows.

Estate	500	
Claims	300	400
Division of estate	150	350

Is this division made according to the contested-garment principle? If not, divide the estate according to the contested-garment principle.

6. A man dies, leaving an estate worth 200 units. The deceased has two creditors; one's claim is 100 units, the other's, 150 units. What should

the return on their claims be according to the contested-garment principle?

7. An estate is divided as follows; check whether the division is made according to the contested-garment principle.

Estate	400	
Claims	200	350
Division of estate	125	275

4.4 A PHYSICAL INTERPRETATION OF THE CONTESTED-GARMENT PRINCIPLE

In this section, we construct a set of vessels that imitate the shares of the creditors according to the contested-garment principle, following Kaminski.[7] Consider, for example, an estate with two claims: 100 and 200. Then imagine two vessels of differing sizes, representing these two claims, into which we pour fluid representing the estate. As shown in Diagrams 1–4, each vessel is composed of two parts connected by a narrow neck. The volume of each part of a vessel is equal to half the claim of the corresponding creditor. The two vessels are connected by a narrow pipe. We assume that the volumes of the necks and the pipe are negligible and considered zero. They serve merely to transfer liquid. We take care that the base areas of the two vessels are equal and that their heights are also equal. Since the claims d_1 and d_2 in this diagram satisfy $d_1 < d_2$, we achieve equal height by constructing a longer neck for the first vessel.

We represent the estate E as a fluid whose volume is equal to E. We pour this fluid into one of the vessels and note that the fluid will stay in the vessels, because $E \leq d_1 + d_2$.

The fluid (estate) that has been poured into one of the vessels now makes its way through the narrow connecting passage into the

[7] Kaminski, M. M. 2000. "Hydraulic rationing," *Mathematical Social Sciences* 40: 131–55

other vessel, ultimately reaching the same level in the two vessels. This simple physical phenomenon is known as "water seeks its own level." We submit that *the amount of fluid in each of the two vessels will then be precisely what the creditor that corresponds to the vessel is entitled to, under the contested-garment principle.* We shall call this the Rule of Linked Vessels.

Let us look at some specific examples.

1. The estate is 80 and the debts are 100 and 200.

Diagram 1

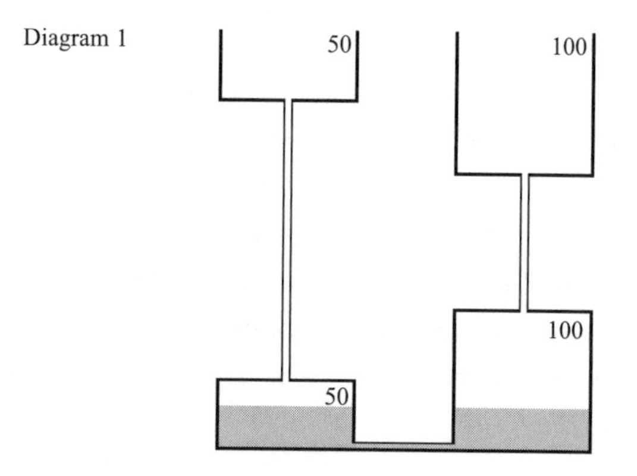

The fluid that has been poured into one of the vessels makes its way through the narrow connecting passage into the other vessel, ultimately reaching the same level in the two vessels.

If we divide the estate of 80 between two creditors of 100 and 200 according to the contested-garment principle it will be:

Estate	80	
Claims	100	200
Uncontested sum	0	0
Contested sum	40	40
Division of estate	40	40

This is exactly what is in the two vessels.

2. The estate is 140 and the debts are 100 and 200.

Diagram 2

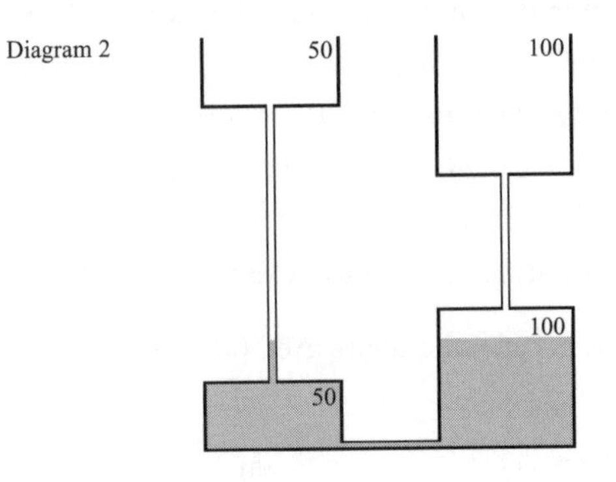

We pour the fluid (estate) and in this case the bottom part of the smaller vessel is full but the fluid does not reach the top part (see diagram). In this case the estate is more than the smaller claim but less than the bigger. When we pour a volume of 140 into the vessels, one vessel will be half filled and the other will be occupied by 90 units of fluid, as we see in the diagram. Computation of the division of E among the creditors is provided below, and we see that *it corresponds exactly to the diagram.*

Estate	140	
Claims	100	200
Uncontested sum	0	40
Contested sum	50	50
Division of estate	50	90

3. The estate is 180 and the debts are 100 and 200.

Diagram 3

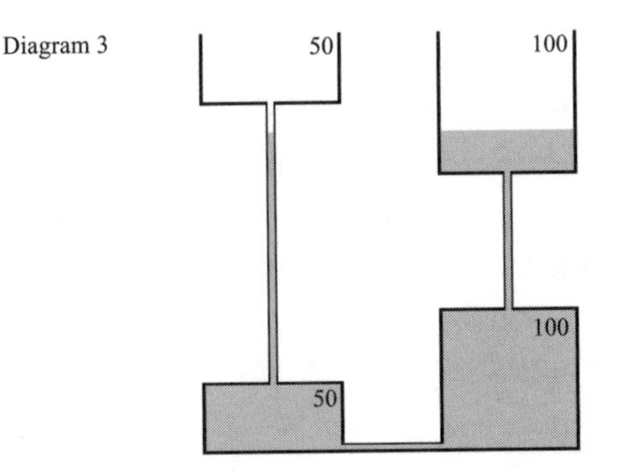

We pour the fluid and in this case, too, the bottom part of the smaller vessel is full but the fluid does not reach the top part. The estate is more than the smaller claim but still less than the bigger one. Here, the fluid occupies only half of the small vessel and 130 units of the other vessel (for a total of 180 units).

The division according to the contested-garment principle *corresponds to the diagram* as the following calculation shows.

Estate	180	
Claims	100	200
Uncontested sum	0	80
Contested sum	50	50
Division of estate	50	130

4. The estate is 240 and the debts are 100 and 200.

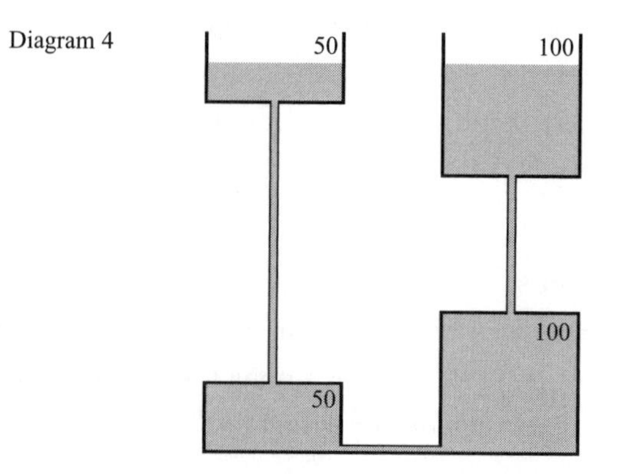

In this case the fluid reaches the top part of both vessels. The sum of the debts is equal to 300 and the estate is equal to 240. There is a shortage of 60 units of fluid which are represented as empty parts of 30 units in each vessel. This shows that both upper halves of the vessels will be filled with fluid.

Computation of the division of the estate in accordance with the contested-garment principle, shown below, *corresponds exactly to the diagram.*

Estate	240	
Claims	100	200
Uncontested sum	40	140
Contested sum	30	30
Division of estate	70	170

These examples illustrate the fact that *for two creditors the construction of the vessels corresponds exactly to the division of an estate E in accordance with the contested-garment principle between two creditors whose claims d_1 and d_2 satisfy $d_1 + d_2 \geq E$.*

This correspondence works in two directions:

1. If we pour the fluid into the vessels and let it seek its own level, the amount of fluid in each vessel will be equal to the amount of fluid prescribed by the contested-garment principle.
2. If we disconnect the vessels and pour into each separately the amount of fluid prescribed by the contested-garment principle and then reconnect the vessels, the fluid will not flow from one vessel into the other, since it will already have reached the same height in both vessels.

4.5 EXERCISES

1. An important way of dividing an estate E among n creditors who claim $d_1, d_2, ..., d_n$ is to divide E among the creditors in proportion to their debts; namely, creditor i will get

$$\frac{d_i}{d_1 + d_2 + ... + d_n} \cdot E$$

Describe a set of vessels and their links that illustrate such a division.

2. A company is owned by three shareholders. The first shareholder owns preferred shares whose nominal value total d_1 and the other two own regular shares whose nominal values are d_2 and d_3. In case of bankruptcy, worth E of the company is distributed to the owners

according to the following rule. First, the first owner gets the nominal value of his shares, as long as $E > d_1$. Otherwise, he gets E. The rest, if any remains, is distributed to creditors 2 and 3 in proportion to their shares d_2 and d_3. Construct vessels that demonstrate how any E satisfying $E \leq d_1 + d_2 + d_3$ is divided.

3. Given an estate E and creditors claiming d_1 and d_2, $d_1 + d_2 \geq E$. Prove that the vessel construction always yields the same division as the contested-garment principle. Hint: We provided four examples above. Construct a general proof for the four examples.

4.6 A BANKRUPTCY PROBLEM FROM THE TALMUD

The Mishna tells of a man with three wives who in their marriage contracts are bequeathed sums of 100, 200, and 300 dinars, respectively. According to the law, these sums are to be paid out to the women when their husband dies. Unfortunately, the husband dies and it turns out that his estate totals less than 600. How should the estate be divided among the widows? The Mishna of Rabbi Nathan discusses three cases:

(i) The estate is 100;
(ii) The estate is 200;
(iii) The estate is 300.

His ruling is presented in the following table:

Claims \ Estate	100	200	300
100	33 1/3	50	50
200	33 1/3	75	100
300	33 1/3	75	150

According to the table, there is equal division among the widows when the estate is 100, there is proportional division among the widows when the estate is 300, but the division is by no means clear when the estate is 200: 50 units to the widow with the marriage contract for 100 and 75 units to each of the other two widows.

Let us consider, for example, the division of the estate among the widows in the second case, where the estate is 200.

Claims \ Estate	200
100	50
200	75
300	75

Let us choose any two widows: the first and the third, for example. The two together get 125 from Rabbi Nathan. What happens if they divide this sum according to the contested-garment principle?

Estate	125	
Claims	100	300
Uncontested division	0	25
Equal division of remainder	50	50
	50	75

According to the contested-garment principle, the claimant to 100 dinars should get 50 and the claimant to 300 dinars should get 75. *Those are the precise amounts Rabbi Nathan specified for the widows!*

Now let us check the division of the estate between the widows with marriage contracts of 200 and 300. The two together get 150 from Rabbi Nathan. According to the contested-garment principle:

Estate	150	
Claims	200	300
Uncontested division	0	0
Equal division of remainder	75	75
	75	75

According to the contested-garment principle, they should each get 75. *Those are the precise amounts Rabbi Nathan specified for the widows!*

A similar calculation shows that if the contested-garment principle is applied to the amount that the widows with marriage contracts for 100 and 200 received together, then they get the precise amounts Rabbi Nathan specified for them (verify this!).

We showed that the division of the estate (50, 75, 75) is *consistent with the contested-garment principle*. Any two widows who share the amount distributed to them in accordance with the contested-garment principle will discover that they get precisely what Rabbi Nathan gave them to begin with.

The remaining cases presented in the table (p. 177) are also consistent with this principle. (In the exercises below you will be asked to verify this.)

It will be proved in Section 4.8 that these are the only numbers consistent with the contested-garment principle. Suppose someone proposes to divide an estate of 200 as (40, 60, 100). Let us check what amount is received by the first widow and the second widow. The two together get 100. Suppose now that the widows are strong believers in the contested-garment principle. Together they received 100. One claims 100 and the other claims 200 units. How should they divide the money that Rabbi Nathan allocated to them? According to the contested-garment principle, they ought to get the following amounts:

Estate	100	
Claims	100	200
Uncontested division	0	0
Equal division of remainder	50	50
	50	50

According to the contested-garment principle, they should get (50,50). Therefore, the first widow will not agree to the proposal above and she will ask for more. In other words, the proposal is *not consistent with the contested-garment principle.*

Let us now check the division of the estate between the second widow and the third widow. According to the proposal above, the two together get 160. According to the contested-garment principle, they ought to get the following amounts:

Estate	160	
Claims	200	300
Uncontested division	0	0
Equal division of remainder	80	80
	—	—
	80	80

Thus, according to the contested-garment principle, they ought to get (80,80). In this case, therefore, the third widow will not agree to the proposed sum and she will oppose it. Thus the widows will oppose the proposal and it will not be implemented. Every time there is a proposal to divide the estate differently than (50,75,75), there will be at least one pair of widows who will find the proposal inconsistent with the contested-garment principle. Only the division (50,75,75) is consistent with the contested-garment principle for each of the three pairs of widows.

4.7 EXERCISES

Note to Exercises 3, 4, 5, and 9: To verify that a solution is consistent with the contested-garment principle, one has to check all the pairs. To conclude that the solution is not consistent with the contested-garment principle, it is enough to find one pair for which the solution is not consistent.

1. An estate is worth 100 units and the claims are 100, 200, and 300 units. Check whether the decision of Rabbi Nathan for each pair of widows is consistent with the contested-garment principle.

2. An estate is worth 300 units and the claims are 100, 200, and 300 units. Check whether the decision of Rabbi Nathan for each pair of widows is consistent with the contested-garment principle.

3. An estate is worth 300 units and the claims are 100, 200, and 300 units. There is a proposal to divide it (80, 90, 130) between three widows. Check whether the sum that every pair of widows receives is different from the sum consistent with the contested-garment principle. (See note above.)

4. An estate is worth 300 units and the claims are 100, 200, and 300 units. There is a proposal to divide it (80, 100, 120) between three widows who claim 100, 200, and 300 units. Check whether there is a pair of widows for whom there is no difference between dividing the estate according to this proposal and dividing the estate according to the contested-garment principle. (See note above.)

5. An estate is worth 200 units and the claims are 100, 200, and 300 units. There is a proposal to divide it (50, 70, 80) between three widows who claim 100, 200, and 300 units. In this case there is only one pair of widows who will oppose the proposal. Which pair is it? (See note above.)

6. A man with an estate worth 400 units goes bankrupt. There are three creditors with claims of 150, 200, and 350 units, respectively. There is a proposal to divide the estate (75, 100, 225) between the creditors. Check whether this proposal is consistent with the contested-garment principle for every pair of creditors.

7. A man with an estate worth 120 units goes bankrupt. There are three creditors with claims of 50, 90, and 130 units, respectively. Is the proposal to divide the estate (25, 45, 50) consistent with the contested-garment principle for every pair of creditors?

8. A man with an estate worth 500 units goes bankrupt. There are three creditors with claims of 150, 250, and 300 units, respectively. The division of the estate is (100, 150, 250). In this case one pair of creditors will get their share of the division according to the contested-garment principle. Which pair is it?

9. A man with an estate worth 200 units goes bankrupt. There are four creditors with claims of 50, 100, 150, and 200 units, respectively. Check whether the proposal to divide the estate $(25, 50, 62\frac{1}{2}, 62\frac{1}{2})$ is consistent with the contested-garment principle for every pair of creditors. (See note above.)

10. A man with an estate of 500 units goes bankrupt. There are four creditors with claims of 100, 150, 250, and 350 units, respectively. Is the proposal to divide the estate $(75, 125, 150, 150)$ consistent with the contested-garment principle for every pair of creditors?

4.8 EXISTENCE AND UNIQUENESS

In the previous sections we studied a specific example from the Talmud and learned that it is a solution that is consistent with the contested-garment principle. Three questions now come to mind.

1. Does there always exist a solution that is consistent with the contested-garment principle? For example, perhaps there is an estate that is bankrupt and its worth has to be shared by 15 creditors whose claims are such that no matter how they share the estate, there will always be two creditors who will find out that what they were offered does not satisfy the contested-garment principle.

2. Is the solution always unique? For example, perhaps there is an eight-person bankruptcy case in which there are two ways to share the estate and both are consistent with the contested-garment principle.

3. What is the solution? Take a five-person bankruptcy situation, with an estate and debts of a given size. How can we find exactly what share each creditor should get that is consistent with the contested-garment principle?

In this section we shall answer questions 1 and 2 affirmatively. The last question will be addressed in Section 4.9.

Theorem:

For any number of claimants and an estate in a bankruptcy situation, there always exists a share of the bankrupt estate that is consistent with the contested-garment principle.

Proof: Let E be an estate and let $d_1, d_2, ..., d_n$ be the non-negative claims against the estate demanded by creditors $1, 2, ..., n$. To be a bankruptcy situation, it must be that

$$E \leq d_1 + d_2 + ... + d_n.$$

To see this, construct n vessels as described in Section 4.4 and connect them as shown in Diagram 5 (done for the case $n = 3$).

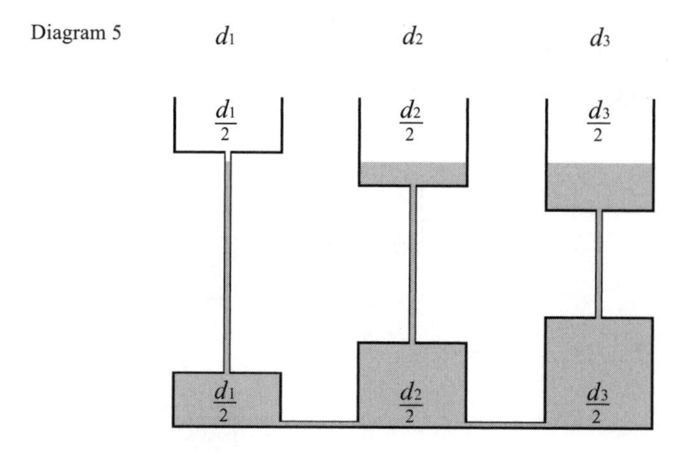

Make sure that the heights as well as the base areas of the three vessels are equal. Now pour E units of fluid into the vessels. The fluid will not overflow the vessels, because $E \leq d_1 + d_2 + ... + d_n$. Let the fluid settle according to the law of "water seeks its own level." Disconnect the pipes connecting the vessels. You get the separate vessels as shown in Diagram 6.

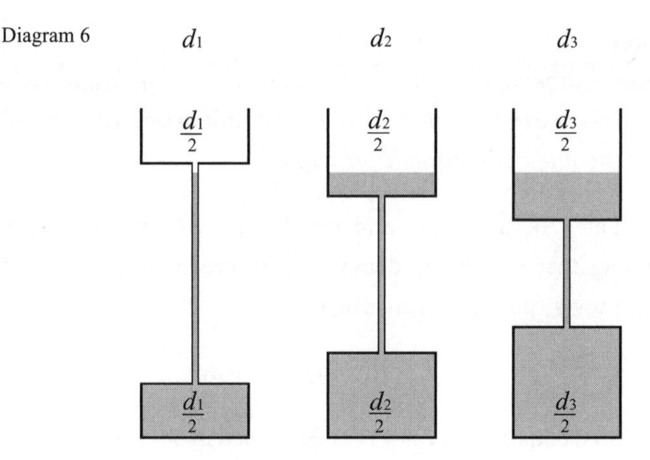

We claim that the amount of fluid in each vessel represents the share of each corresponding creditor. Note that the fluid reaches the same height in all three vessels.

Take any two vessels i and j and connect them (Diagram 7).

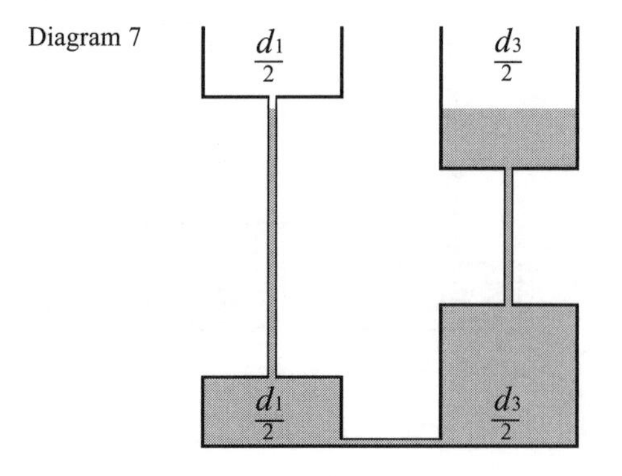

Notice that the fluid does not flow from one vessel into the other, because the height of the fluid in both vessels is already the same. Thus, the fluid in vessels i and j obeys the Rule of Linked Vessels. This proves that the share that we propose is consistent with the contested-garment principle.

Theorem:

There is only one way to share the estate E with creditors $d_1, d_2, ..., d_n$ that is consistent with the contested-garment principle and it is the one described in the previous theorem.

Proof: Let $e_1, e_2, ..., e_n$ be a share of E that is consistent with the contested-garment principle. Consider the vessels as before but do not connect them as of yet (Diagram 8).

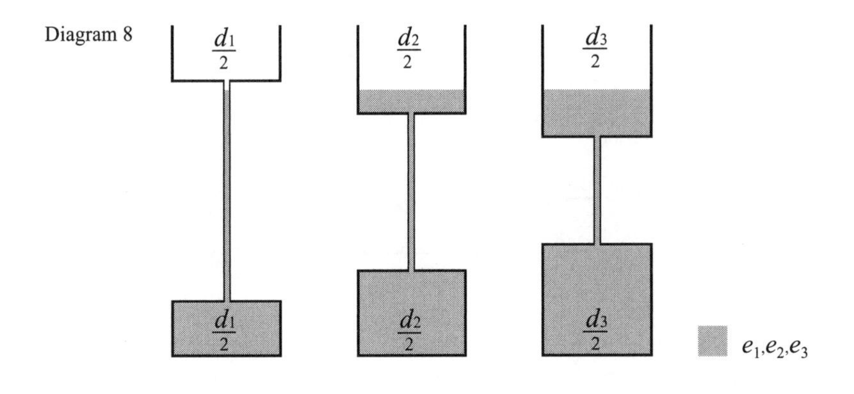

Pour amounts of fluid $e_1, e_2, ..., e_n$ into vessels $1, 2, ..., n$. Take any two vessels i and j and connect them (Diagram 9).

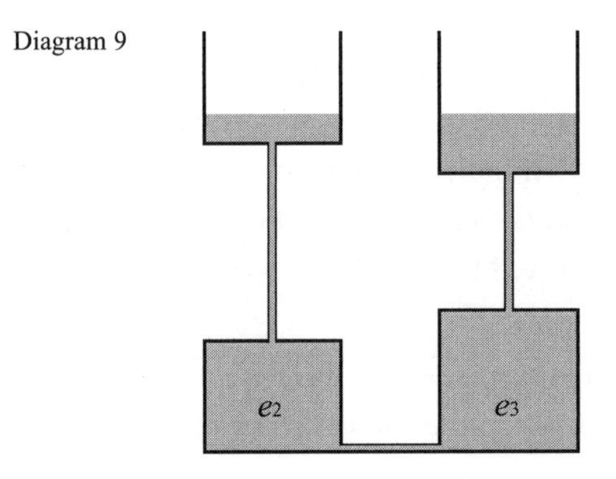

The fluid will be at the same level because the solution $(e_1, e_2, ..., e_n)$ is consistent with the contested-garment principle. This is true for every pair of vessels, so the fluid is at the same height in all of them. (Explain.) Now connect all the vessels and you see that no fluid will flow from one vessel into the others (Diagram 10 for the case $n = 3$).

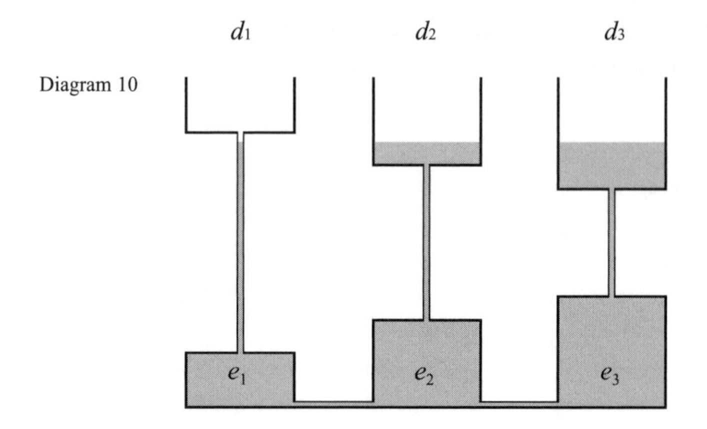

This shows that a consistent contested-garment principle must be as described in the previous theorem.

4.9 DIVISIONS CONSISTENT WITH THE CONTESTED-GARMENT PRINCIPLE

At this point one may ask what the recommended division should be if the estate is not necessarily worth 100, 200, or 300 units. In this section we shall present a law for the division of an estate that is worth less than the sum of its claims. We shall discuss the case of three widows with marriage contracts of 100, 200, and 300 when the values of the estate are different from these. From the rule that we shall establish it will be clear how to extend it to any claims and any number of creditors.

The following table (Table 1) describes three cases where the law is enforced, when the estate has an upper bound of 300.

Table 1:

Claims \ Estate		150		250		300
100	α	50	50	50	50	50
200	α	50	50+β	100	100	100
300	α	50	50+β	100	100+γ	150

Explanation: When the estate is small, it is divided equally among the widows (first column from left). Every unit of money that comes from the estate is divided equally among the widows. That is, it is divided equally among them until the first widow obtains half of her marriage contract (second column from left). At this stage the estate is 150.

From this stage on, each additional unit of money is divided equally between the second widow and the third widow (third column from left). That is, each additional unit is divided equally between the second widow and the third widow until the second widow obtains half of her marriage contract. At this stage the estate is 250. From this stage on, each additional unit of money is given to the third widow only (fifth column from left). That is, each additional unit is given to the third widow until she obtains half of her marriage contract. The estate at this stage is 300.

The following table describes the enforcement of the law when the estate exceeds 300 units, but does not exceed 600 units.

Table 2:

Claims \ Estate	300		350		450		600
100	50	50	50	50	50	100–α	100
200	100	100	100	150–β	150	200–α	200
300	150	200–γ	200	250–β	250	300–α	300

Explanation: In this case we examine the losses. When the estate is 600 units (or more), there is no problem in dividing it; each widow obtains her marriage contract (first column from right). When the estate is under 600 units, the widows incur equal losses (second column from right). That is, they incur equal losses until the first widow loses half of her marriage contract. At this stage the estate is 450 units. From this stage on, each additional loss is divided equally between the second widow and the third widow only (fourth column from right), until the second widow loses half of her marriage contract. The estate at this stage is 350 units. From this stage on, only the third widow incurs losses (sixth column from right), until she loses half of her marriage contract and the estate is 300 units.

Now we shall check the second column in Table 2, when $\gamma = 15$.

Claims \ Estate	335
100	50
200	100
300	185

We conclude by verifying that this division indeed obeys the contested-garment principle.

150		235		285	
100	200	100	300	200	300
0	50	0	135	0	85
50	50	50	50	100	100
—	—	—	—	—	—
50	100	50	185	100	185

This law can easily be generalized to cases where the claims are different and the number of claimants is greater.

Example 1

There are four creditors with claims of 120, 140, 200, and 250 units of money, respectively. The estate that is divided to cover the debt is worth only 300 units. How should they divide the estate according to the law described above?

Solution: The total amount of claims in this case is 710 units. The estate supposed to cover the debt is worth 300 units; i.e., it is less than half the total amount of claims. It will be helpful, therefore, to consult Table 1.

Let us complete the table as described above, until we exceed the amount of 300 units.

Claims \ Estate	240	270	330
120	60	60	60
140	60	70	70
200	60	70	100
250	60	70	100

We exceed the amout of 300 units when we divide each additional unit equally between the two last claimants. Thus, we shall deduct the surplus amount, when it is evenly divided between the two.

The division obtained is (60, 70, 85, 85).

Exercise: Check whether the amount received by the first claimant and the third claimant in this division is consistent with the contested-garment principle.

Example 2

There are four creditors with claims of 120, 140, 200, and 250 units of money, respectively. The estate to be divided is 420 units. How can the estate be divided in a way that is consistent with the contested-garment principle?

Solution: In this case, the estate to be divided is greater than half the amount of the debts ($710 \div 2 = 355$). We are concerned, therefore, with the losses and it will be helpful to consult Table 2, which we shall complete from right to left:

Claims \ Estate	380	440	470	710
120	60	60	60	120
140	70	70	80	140
200	100	130	140	200
250	150	180	190	250

In the last column we have obtained less than the total amount at our disposal to divide; i.e., we have deducted too much from the third claimant and the fourth claimant. We must divide 420; hence it is necessary to add 40 units, which are divided equally between the last two claimants. The requested division, therefore, is (60, 70, 120, 170).

For example, let us check whether the second claimant and the fourth claimant have received amounts consistent with the contested-garment principle.

240	
140	250
0	100
70	70
—	—
70	170

Thus, our check shows consistency with the contested-garment principle.

Summary: In this section we introduced a procedure for the division of an estate among creditors. Implementation of this procedure requires partial completion of a table – in terms of profits, if the estate is less than half the amount of the claims, and in terms of losses, if the estate is greater than half the amount of the claims. One completes the table, until one gets the correct division. The reader can ascertain that the procedure described above imitates the liquid poured into the vessels. This proves the following theorem:

Theorem:
The procedure for the division of an estate described above results in the end in a division consistent with the contested-garment principle for every pair of creditors. By any other division there will be at least one pair of creditors for whom the amounts received are not consistent with the contested-garment principle.

On the basis of this theorem, every outcome of this procedure is *contested-garment-consistent.*

4.10 EXERCISES

1. A man dies, leaving an estate worth 500 units of money. The deceased has three widows with marriage contracts of 100, 200, and 300 units, respectively. Divide the estate among the widows, such that the division is contested-garment-consistent.

2. Divide an estate worth 300 units among three widows with claims of 50, 100, and 200 units, respectively, such that the division is contested-garment-consistent.

3. Divide an estate worth 230 units among four widows with claims of 50, 100, 150, and 200 units, respectively, such that the division is contested-garment-consistent.

4. Divide an estate worth 350 units among four widows with claims of 80, 120, 160, and 200 units, respectively, such that the division is contested-garment-consistent.

5. Divide an estate worth 800 units of money among six widows with claims of 50, 100, 150, 200, 250, and 300 units, respectively, such that the division is contested-garment-consistent.

6. Divide an estate worth 400 units among five widows with claims of 70, 100, 160, 220, and 300 units, respectively, such that the division is contested-garment-consistent.

7. Check whether the (25, 75, 125, 175) division of an estate worth 400 units among four widows with claims of 50, 100, 150, and 200 units, respectively, is contested-garment-consistent.

8. Check whether the (50, 100, 150, 200, 200) division of an estate worth 700 units among five widows with claims of 75, 125, 200, 250, and 300 units, respectively, is contested-garment-consistent.

9. The following table presents divisions of an estate in various amounts for four creditors. (The upper row represents the different estates and the left-hand column represents the different claims.) Check whether all the divisions in the table are contested-garment-consistent. Indicate which divisions are not contested-garment-consistent.

Claims \ Estate	100	150	200	300	400
50	25	37.5	25	25	25
100	25	37.5	50	50	75
200	25	37.5	62.5	100	150
300	25	37.5	62.5	125	150

4.11 CONSISTENCY

Let us return to Example 2 in Section 4.9. This example involves four creditors with claims and a division of the estate among them as follows.

	420
120	60
140	70
200	120
250	170

The question is, assuming the amount received by three of the four claimants (say, the first, third, and fourth, who together get 350 units) is divided according to the contested-garment principle, is the same division obtained as when the estate is divided among three of the four claimants?

The answer to this question is positive and can be proved in two ways.

First Proof: The part of the estate paid by the three claimants is the same as in the original problem and in the three-person problem, namely, 350 units. The division of this sum is consistent with the contested-garment principle for any pair of players, and, in particular, for any pair in the three-person problem. Thus, the solution for the four-person problem, restricted to the three-person problem, is indeed consistent with the contested-garment principle.

Second Proof: We construct the appropriate table until we have two adjacent columns: one with the estate above 350 units and one with the estate below 350 units.

	310	390	570
120	60	60	120
200	100	140	200
250	150	190	250

In this range deduction takes place only between the last two claimants. From 350 units we still have to deduct 40 units, to be divided equally between these claimants. We get the division (60, 120, 170), which is exactly what all three creditors received in the four-person problem.

The above example is a special case of the following theorem:

Theorem:
If a set of creditors divides an estate according to the contested-garment principle, then each subset that divides the amount that its members obtained in the original division, while respecting the original claims and according to the contested-garment principle, will get precisely the same division that they obtained in the original division.

We can summarize the theorem as follows:

A division according to the contested-garment principle is a division that is consistent for any number of its participants (and not just for any two participants).

4.12 EXERCISES

1. (1) Divide an estate worth 550 units of money among four creditors with claims of 50, 150, 200, and 300 units, respectively, according to the contested-garment principle.
   (2) Check whether the total amount received by the claimants to 50, 200, and 300 units will be divided among them in the same way if it is divided among the three of them according to the contested-garment principle.
2. (1) Divide an estate worth 400 units among six creditors with claims of 50, 80, 100, 140, 200, and 250 units, respectively, according to the contested-garment principle.
   (2) Check whether the total amount received by the claimants to 80, 140, 200, and 250 units will be divided among them in the same way if it is divided among the four of them according to the contested-garment principle.
3. (1) Divide an estate worth 900 units among six creditors with claims of 100, 150, 200, 260, 300, and 320 units, respectively, according to the contested-garment principle.
   (2) Check whether the total amount received by the claimants to 100, 200, and 300 units will be divided among them in the same way if it is divided among the three of them according to the contested-garment principle.

4.13 RIF'S LAW OF DIVISION

The Rif (Rabbi Yitzhak Alfasi) proposed another law of division, later adopted by Rambam (Rabbi Moshe Ben Maimon). According to this law, every unit of money is divided equally among all claimants, until the claimant with the smallest claim gets his full amount. Each additional unit of money is divided equally among the remaining creditors, until the claimant with the smallest claim at this stage gets his full amount, and so on.

Example:

Claims \ Estate		300		500		600
100	α	100	100	100	100	100
200	α	100	100+β	200	200	200
300	α	100	100+β	200	200+γ	300

Explanation: Equal division occurs until the first claimant receives her claim (the 300-column). Then, the other claimants receive additional equal amounts until the second claimant receives her claim (the 500-column). At that stage, the last claimant receives the remainder of the estate, but not more than his claim (the 600-column).

Suppose now that the estate is 350 units. To divide it, we construct the above table up to and inclusive of the 500-column (explain) and we see that the division must be (100, 125, 125).

Is Rif's law of division consistent? That is, will every subset of claimants that divides the total received by the claimants in the original division according to Rif's law, get the same amounts? Let us check how much the first claimant and the third claimant get when the estate is 350. The two together got 225. We shall divide this sum between them according to Rif's law.

Claims \ Estate	200	225
100	100	100
300	100	125

Explanation: First, we divided 100 units for each of the two claimants. The first got his full amount. The rest was given to the second claimant.

We see that in this case there is consistency between the first and third creditors.

Exercise: Check whether there is consistency between the first and second creditors, and between the second and third creditors.

It can be proved that Rif's law of division is indeed consistent; i.e., for any subset, if we distribute among its members the total payoffs they received together in the original division according to Rif's law, the same division will be obtained.

4.14 EXERCISES

1. Divide an estate worth 275 units among four creditors with claims of 50, 100, 150, and 200 units, respectively, according to Rif's law.

2. Divide an estate worth 400 units among four creditors with claims of 50, 100, 150, and 200 units, respectively, according to Rif's law.

3. (1) Divide an estate worth 790 units among five creditors with claims of 100, 150, 200, 250, and 300 units, respectively, according to Rif's law.
   (2) Check whether the division according to Rif's law is consistent, say, for a group of creditors with claims of 150, 250, and 300 units, respectively.

4. (1) Divide an estate worth 400 units among five creditors with claims of 40, 60, 80, 120, and 150 units, respectively, according to Rif's law.
   (2) Check whether the division according to Rif's law is consistent, say, for a group of creditors with claims of 40, 60, 80, and 150 units, respectively.

4.15 PROPORTIONAL DIVISION

In the world of finance it is customary to divide an estate in proportion to the investments.

Example:

Four partners founded a company that later closed due to financial difficulties. We divide its market value – $555,000 – among the partners proportionally to their shares in the company, which are 40, 60, 120, and 150, respectively.

When the total number of shares in the company is $40 + 60 + 120 + 150 = 370$:

The first gets: $\frac{555 \cdot 40}{370} = 60$
The second gets: $\frac{555 \cdot 60}{370} = 90$
The third gets: $\frac{555 \cdot 120}{370} = 180$
The fourth gets: $\frac{555 \cdot 150}{370} = 225$

Question: Is the law of proportional division consistent?

Answer: Let's check for the first three shareholders, who received $60 + 90 + 180 = 330$. We divide this amount proportionally among them. When their shares are $40 + 60 + 120 = 220$, then:

The first gets: $\frac{330 \cdot 40}{220} = 60$

The second gets: $\frac{330 \cdot 60}{220} = 90$

The third gets: $\frac{330 \cdot 120}{220} = 180$

Thus, the amounts the shareholders obtain are precisely those they got in the original division.

It is easy to prove that the proportional division is also a consistent solution. Any subset of players who examine the amounts distributed to them will find them proportional to their claims.

4.16 O'NEILL'S LAW OF DIVISION

O'Neill presents another interesting law.[8] Consider, for example, a case where the estate is worth 250 units and the claims are 100, 200, and 300 units, respectively. The creditors rush to the bank or to wherever the estate is disbursed. The first to arrive gets his claim

[8] O'Neill, B. 1982. "A problem of rights arbitration from the Talmud," *Mathematical Social Sciences* 2: 345–71

in full, because no other claims have been presented. The second to arrive gets his claim in full or in part, depending on the amount of money left over from the first claim, and so on. Every creditor to arrive gets his claim in full or in part, until the estate is depleted. The amount each creditor gets depends, of course, on the order of arrival.

O'Neill's law proposes that, instead of holding a race, each creditor compute all he can get according to the order of his arrival, over all possible orders. The amount each creditor gets in the end will be the average of the amounts received in all possible orders.

The estate is 250:
1 claims 100:
2 claims 200:
3 claims 300:

Order of arrival \ Creditors	1	2	3
123	100	150	0
132	100	0	150
213	50	200	0
231	0	200	50
312	0	0	250
321	0	0	250
	(250,	550,	700):6 = (41⅔, 91⅔, 116⅔)

The final division is the average of the amounts, namely, $\left(41\frac{2}{3}, 91\frac{2}{3}, 116\frac{2}{3}\right)$.

Is O'Neill's law consistent?

Let us consider, say, the first and third creditors. They together received $158\frac{1}{3}$, while their claims are 100 and 300, respectively. The division according to O'Neill's law will be as follows.

The estate is $158\frac{1}{3}$:
1 claims 100:
2 claims 300:

Order of arrival \ Creditors	1	2
12	100	$58\frac{1}{3}$
21	0	$158\frac{1}{3}$
	$(100, 216\frac{2}{3}):2 = (50, 108\frac{1}{3})$	

According to the law the creditors will get $(50, 108\frac{1}{3})$, which is not the original division. We see that this law of division does not satisfy the consistency property.

Every bankruptcy problem of the kind we have discussed up until now can be translated to an $(N; v)$ game where N is the set of creditors and the coalition function is defined as:

$v(S) =$ [estate minus amount of claims of creditors who are not in $S]_+$

Explanation: The amount due to the individuals in S without any division is the sum that is left over from the estate after the creditors who are not in S receive their claims in full. Thus the creditors in S can guarantee themselves this amount. If the difference between the two amounts is negative, then we set $v(S) = 0$, and that is the meaning of the plus sign in the formula above.

Let us now translate the example above to an $(N; v)$ game: the estate totals 250 units and creditors 1, 2, and 3 claim 100, 200, and 300 units, respectively.

$$N = \{1, 2, 3\}$$

$$v(1) = [250 - (200 + 300)]_+ = 0$$
$$v(2) = [250 - (100 + 300)]_+ = 0$$
$$v(3) = [250 - (100 + 200)]_+ = 0$$
$$v(1, 2) = [250 - 300]_+ = 0$$
$$v(1, 3) = [250 - 200]_+ = 50$$
$$v(2, 3) = [250 - 100]_+ = 150$$
$$v(1, 2, 3) = [250 - 0]_+ = 250$$
$$v(\emptyset) = [250 - 250]_+ = 0$$

The game can be presented in the following figure:

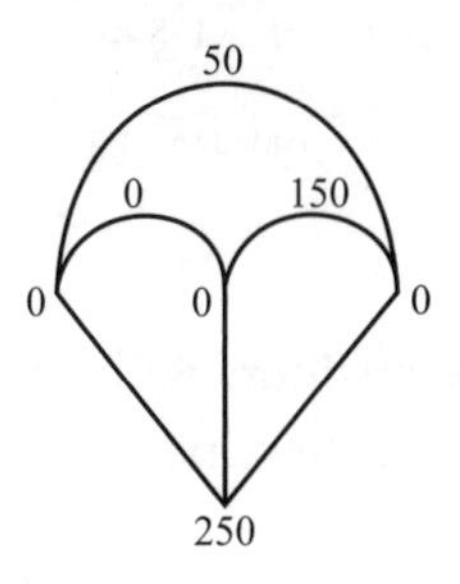

We calculate the Shapley value of the game.

Orders \ Creditors	1	2	3
123	0	0	250
132	0	200	50
213	0	0	250
231	100	0	150
312	50	200	0
321	100	150	0
	(250,	550,	700):6 = $(41\frac{2}{3}, 91\frac{2}{3}, 116\frac{2}{3})$

We see that the Shapley value of this game is precisely O'Neill's solution.

4.17 EXERCISES

1. Divide an estate worth 400 units among three creditors with claims of 100, 200, and 300 units, respectively, according to O'Neill's law.

2. Divide an estate worth 500 units among four creditors with claims of 100, 150, 200, and 250 units, respectively, according to O'Neill's law.

3. There is a bankruptcy problem in which an estate is worth 500 units and the claims are 100, 300, and 400 units, respectively. Translate the problem to a coalition game and calculate the Shapley value of the game. Show that the O'Neill procedure leads to the same division.

4. There is a bankruptcy problem in which an estate is worth 700 units and the claims are 200, 250, 300, and 400 units, respectively. Translate the problem to a coalition game and calculate the Shapley value of the game. Show that the O'Neill procedure leads to the same division.

5. (1) Divide an estate worth 200 units among four creditors with claims of 50, 100, 150, and 200 units, respectively, according to O'Neill's law.
   (2) Show that the law is not consistent by checking a pair of creditors.

6. (1) Divide an estate worth 300 units among four creditors with claims of 80, 120, 200, and 280 units, respectively, according to O'Neill's law.
   (2) Translate the problem to a coalition game.
   (3) Show that O'Neill's law is not consistent by checking three creditors.

4.18 DISCUSSION

In this chapter we presented four different laws for dividing an estate among creditors when the total amount of the claims against the estate exceeds the value of the estate. Note that some of them apply to various situations in real life. For example, the proportional law applies when a company of shareholders goes bankrupt. O'Neill's "running to the bank" solution, which is also the Shapley value of an appropriate coalition function, can be understood as an a priori expectation in those cases where the players actually run to the bank and there is no way of telling in advance in what order they will arrive there. The Talmudic law of Rabbi Nathan can be considered desirable when the players want to share equally the contested part of the debts.

Can we say which solution is superior to the others? Obviously not, because each of them is considered better suited to a particular case.

Since we cannot say that any one solution is absolutely superior, which solution should we recommend when a new real-life situation arises? Even if each solution sheds light on different aspects of the case, we usually have to decide on a single solution. How does one make this choice? That is, what criteria should guide one's choice in preferring any one solution to the others in a given real-life case? We can only provide guidelines:

(a) Look at the axioms and properties that characterize each solution and see which axioms better fit the reality. For example, the requirement of consistency is sometimes appealing, and it is this requirement that gives rise to the proportional solution, the Talmudic solution of Rabbi Nathan, and others.

(b) Look at the behavior of the players in real life. For example, perhaps they do "run to the bank," in which case O'Neill's solution, which is also the Shapley value, yields an a priori expectation on the final settlement.

(c) In complex situations, offer the players a simpler problem to which they can suggest an intelligent solution, say, a two-person case, and try to learn from their choice what aspects of the simpler problem they focus on. Then generalize to the more complex real-life case.

Broadly speaking, all the chapters in this book represent attempts at reaching a decision in a conflict situation and in each of them we show the difficulties when trying to define a "superior" solution. The first chapter on matching presents a "weak" condition of stability, which nevertheless yields many matchings. One of them is best for the men and another is best for the women. The second chapter tries to reach a decision by voting and we saw that a fair voting rule is not always possible. The third chapter has probably the most successful solution. It provides a solution for an unbiased arbitrator, by supplying axioms that seem fair. However, somewhat different axioms, not covered in this book, yield different solutions. Finally, the fourth chapter, which considers the case of bankruptcy conflicts, shows that even in this simple case a superior solution cannot be defined.

In conclusion, we see that various solutions are well tailored to many real situations, but there is no single solution that fits all situations. Each solution sheds some light on the reality.

4.19 REVIEW EXERCISES

1. A man dies, leaving an estate worth 500 units of money. The deceased has two creditors; one claims 350 units and the other claims 300 units. How will the estate be divided between them according to the contested-garment principle, Rif's law of division, proportional division, and O'Neill's law of division?

2. A man with an estate worth 1000 units goes bankrupt. The bankrupt man has four creditors with claims of 200, 300, 400, and 500 units of money, respectively. Divide the estate among the creditors according to the contested-garment principle, Rif's law of division, proportional division, and O'Neill's law of division.

3. (1) Divide an estate worth 800 units among six creditors with claims of 50, 100, 150, 200, 250, and 300 units, respectively, according to the contested-garment principle, Rif's law of division, and proportional division.
   (2) Check whether the total amount received by the creditors with claims of 50, 150, 250, and 300 units will be divided according to the divisions specified in 3(1).
4. (1) Divide an estate worth 700 units among four creditors with claims of 100, 200, 250, and 350 units of money, respectively, according to the contested-garment principle, Rif's law of division, and proportional division.
   (2) Check whether the division is consistent for a group of creditors with claims of 100, 250, and 350 units, respectively, for the divisions specified in 4(1).

5. There is a bankruptcy problem in which an estate is 800 units and the claims are 200, 300, and 400 units, respectively. Translate the problem to a coalition game and calculate the Shapley value of the game. Show that the O'Neill procedure leads to the same division.

Appendix Answers to the Exercises

A.1 CHAPTER 1

1.3

1. i. Stable
1. ii. Unstable

2. i. Unstable
2. ii. Unstable
2. iii. Unstable
2. iv. Stable

3. (1) Stable
3. (2) Unstable (Bc)

4. (1) i. Stable. All men get their first choice.
4. (1) ii. Stable. All men get their second choice and every one of them is the fifth choice of his first choice.
4. (1) iii. Stable. All men and women get their third choice and every move will worsen their situation.
4. (1) iv. Stable. All women get their second choice and every one of them is the fifth choice of her first choice.
4. (2) The preference structure is cyclic, such that there is one move in both the men's and the women's choices, but the move in every case is in the opposite direction. For example, all the men's first choices are women who ranked them fifth and all the men's second choices are women who ranked them fourth, and so on.

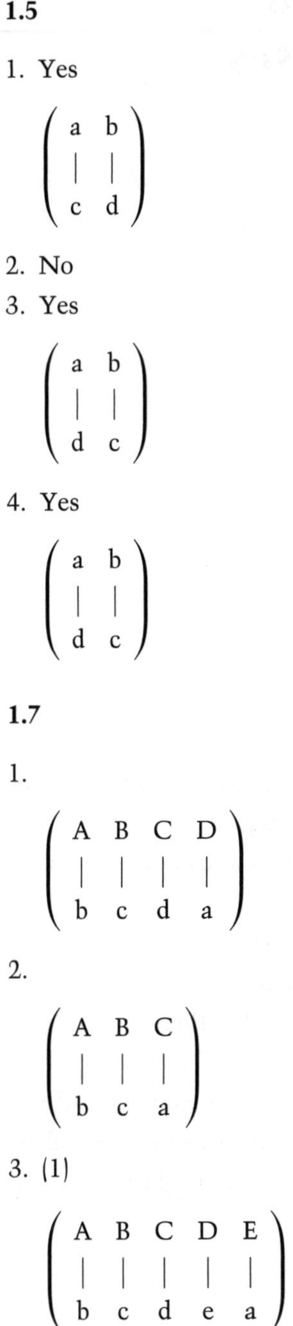

1.5

1. Yes

$$\begin{pmatrix} a & b \\ | & | \\ c & d \end{pmatrix}$$

2. No
3. Yes

$$\begin{pmatrix} a & b \\ | & | \\ d & c \end{pmatrix}$$

4. Yes

$$\begin{pmatrix} a & b \\ | & | \\ d & c \end{pmatrix}$$

1.7

1.

$$\begin{pmatrix} A & B & C & D \\ | & | & | & | \\ b & c & d & a \end{pmatrix}$$

2.

$$\begin{pmatrix} A & B & C \\ | & | & | \\ b & c & a \end{pmatrix}$$

3. (1)

$$\begin{pmatrix} A & B & C & D & E \\ | & | & | & | & | \\ b & c & d & e & a \end{pmatrix}$$

3. (2)

$$\begin{pmatrix} A & B & C & D & E \\ | & | & | & | & | \\ b & c & d & e & a \end{pmatrix}$$

1.11

1.

$$\begin{pmatrix} A & B & C & - \\ | & | & | & | \\ b & c & a & d \end{pmatrix}$$

2. (1)

$$\begin{pmatrix} A & B & C & D & E \\ | & | & | & | & | \\ b & c & a & - & - \end{pmatrix}$$

2. (2)

$$\begin{pmatrix} A & B & C & D & E \\ | & | & | & | & | \\ c & b & a & - & - \end{pmatrix}$$

3.

$$\begin{pmatrix} A & B & C & - \\ | & | & | & | \\ d & c & b & a \end{pmatrix}$$

4. (i) Unstable (Aa)
4. (ii) Stable
4. (iii) Unstable (Ac)
4. (iv) Unstable (Bc)
4. (v) Unstable (Bc)

4. (vi) Stable

No other matching system exists. For a community of 3×3 there are $3! = 6$ matching systems.

5. (1)

$$\begin{pmatrix} A & B & C \\ | & | & | \\ a & c & b \end{pmatrix}$$

5. (2)

$$\begin{pmatrix} A & B & C \\ | & | & | \\ b & c & a \end{pmatrix}$$

6. (1)

$$\begin{pmatrix} A & B & C & - & - \\ | & | & | & | & | \\ b & a & - & d & c \end{pmatrix}$$

6. (2)

$$\begin{pmatrix} A & B & C & - \\ | & | & | & | \\ b & d & a & c \end{pmatrix}$$

7. (1) Male courtship:

$$\begin{pmatrix} A & B & C & D & - & - \\ | & | & | & | & | & \\ c & d & a & - & b & e \end{pmatrix}$$

We get the same matching system in female courtship.

7. (2) We get the same matching system in male courtship as in 7 (1).
We get the same matching system again in female courtship.

8.

$$\begin{pmatrix} A & B & C \\ | & | & | \\ a & c & b \end{pmatrix} \qquad \begin{pmatrix} A & B & C \\ | & | & | \\ c & a & b \end{pmatrix}$$

9.

$$\begin{pmatrix} A & B & C & D \\ | & | & | & | \\ c & d & a & b \end{pmatrix}$$

1.13

1.

$$\begin{pmatrix} A & B & C \\ | & | & | \\ a,c & b,e & i,n,l \\ d & k,g & j,o,h \end{pmatrix}$$

m, f are out.

2.

$$\begin{pmatrix} A & B & C & D & E \\ | & | & | & | & | \\ a,c,h,j & e,g & m,n & p,r & b \\ k,l,t & & & s & \end{pmatrix}$$

d, f, i, o, q are out.

3. (1)

$$\begin{pmatrix} A & B & C & D & E \\ | & | & | & | & | \\ c,e,g & b,d & h & i,j & a \end{pmatrix}$$

(2) The same division is obtained as in 3(1).

1.15

2. (1) Not all women are possible for Mr. d. Ms. C is impossible for Mr. d.

2. (2) The women possible for Mr. a are A and D.

2. (3) Ms. A is possible for a and d.
Ms. C is possible for b and c.
Ms. D is possible for a, c, and d.

3. (1) A is impossible for a.

3. (2) A is possible for b.

3. (3) B is possible for b.

4. (1)

$$\begin{pmatrix} A & B & C & D & E \\ | & | & | & | & | \\ e & b & a & c & d \end{pmatrix}$$

4. (2)

$$\begin{pmatrix} A & B & C & D & E \\ | & | & | & | & | \\ d & e & b & a & c \end{pmatrix}$$

4. (3) No

5. (1)

$$\begin{pmatrix} A & B & C & - \\ | & | & | & | \\ a & d & b & c \end{pmatrix}$$

5. (2) The same matching system is obtained as in 5(1).

6. (1)

$$\begin{pmatrix} A & B & C \\ | & | & | \\ c & b & a \end{pmatrix}$$

(2)

$$\begin{pmatrix} A & B & C \\ | & | & | \\ c & a & b \end{pmatrix}$$

System (1) is not optimal for the men in the original system, because system (2) is better for Mr. a who prefers B to C.

7. (1)

$$\begin{pmatrix} A & B & C & D \\ | & | & | & | \\ b & c & d & a \end{pmatrix}$$

1.17

1. Yes

2. No

   Optimal system for men:

$$\begin{pmatrix} A & B & C & D & - \\ | & | & | & | & | \\ e & c & b & a & d \end{pmatrix}$$

   Optimal system for women:

$$\begin{pmatrix} A & B & C & D & - \\ | & | & | & | & | \\ b & a & c & e & d \end{pmatrix}$$

3. No

4. No

$$\begin{pmatrix} A & B & C & - \\ | & | & | & | \\ a & c & d & b \end{pmatrix}$$

1.19

1. No

2. (1)

$$\begin{pmatrix} A & B & C & D \\ | & | & | & | \\ a & b & c & d \end{pmatrix}$$

2. (2) Yes

3. (1)

$$\begin{pmatrix} A & B & C & D & E \\ | & | & | & | & | \\ - & a & b & c & - \end{pmatrix}$$

3. (2) The women who remain without a mate are A and E. They will remain without a mate in every case.

4.

$$\begin{pmatrix} A & B & C & D \\ | & | & | & | \\ a & d & b & c \end{pmatrix}$$

5. (1) The system obtained in male courtship is

$$\begin{pmatrix} A & B & C \\ | & | & | \\ a & b & c \end{pmatrix}$$

and in female courtship:

$$\begin{pmatrix} A & B & C \\ | & | & | \\ c & b & a \end{pmatrix}$$

5. (2) In male courtship the same system is obtained in every case but in female courtship the following system is also obtained:

$$\begin{pmatrix} A & B & C \\ | & | & | \\ b & c & a \end{pmatrix}$$

6.

$$\begin{pmatrix} A & B & C & - \\ | & | & | & | \\ a & b & c & d \end{pmatrix} \quad \begin{pmatrix} A & B & C & - \\ | & | & | & | \\ d & a & b & c \end{pmatrix} \quad \begin{pmatrix} A & B & C & - \\ | & | & | & | \\ b & a & c & d \end{pmatrix}$$

7. (1)

$$\begin{pmatrix} A & B & C & D & E \\ | & | & | & | & | \\ c & b & e & a & d \end{pmatrix}$$

7. (3) No

A.2 CHAPTER 2

2.3

1. (1) x: return money to students
 y: buy theater tickets
 z: organize end-of-year party

1	**2**	**3**
x	y	z
y	z	y
z	x	x

1. (2)

$$\begin{pmatrix} y \\ z \\ x \end{pmatrix}$$

2. (1) p: pizza
 s: sandwich
 h: hamburger
 b: burrito
 t: taco
 f: falafel

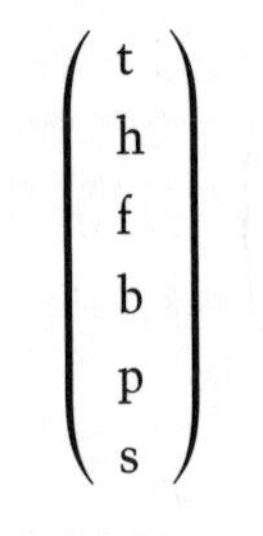

2. (2) Yes

2. (3) The superfluous information, which derives from the transitivity of the preference relation, is $\begin{matrix} z \\ y \end{matrix}$.

3. (1) x: movies
   y: theater
   z: dancing
   p: show
   q: reading
   r: concert
   w: TV

$$\begin{pmatrix} w \\ q \\ r \\ x \\ y \\ z \\ p \end{pmatrix}$$

3. (2) The superfluous information, which derives from the transitivity of the preference relation, is $\begin{matrix} w \\ y \end{matrix}$.

4. Sarah prefers Bach to Chopin.

5. (1) First preference: sunny-side down.

5. (2) There is not enough information to place the soft-boiled egg in the preference order. It is necessary to ask what the preference

is between a soft-boiled egg and a hard-boiled egg and between a soft-boiled egg and an egg sunny-side up.

5. (3) It contradicts assumption A, because by transitivity of the preference relation we get a preference for sunny-side up over sunny-side up.

6. (1) First preference: science fiction.
6. (2) What is preferred: (i) comedy or western?
   (ii) comedy or science-fiction film?
6. (3) There is a contradiction: action film is preferred to horror film, but we derive the opposite from the transitivity of the preference relation. Similarly, horror film is preferred to western, but we derive the opposite from the transitivity of the preference relation.

7. (1) Coffee or cappuccino.
7. (2) Coffee
7. (3) Cappuccino

2.4

I. (1)

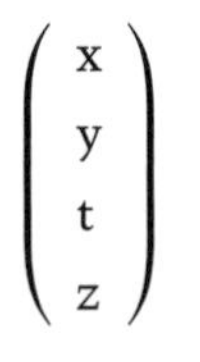

I. (2) No decision

III.

$$\begin{pmatrix} t \\ z \\ y \\ x \end{pmatrix}$$

IV. (1) No decision

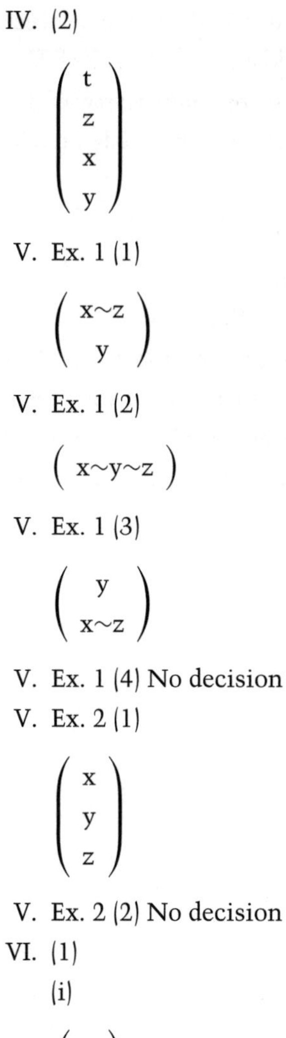

IV. (2)

$$\begin{pmatrix} t \\ z \\ x \\ y \end{pmatrix}$$

V. Ex. 1 (1)

$$\begin{pmatrix} x \sim z \\ y \end{pmatrix}$$

V. Ex. 1 (2)

$$\begin{pmatrix} x \sim y \sim z \end{pmatrix}$$

V. Ex. 1 (3)

$$\begin{pmatrix} y \\ x \sim z \end{pmatrix}$$

V. Ex. 1 (4) No decision

V. Ex. 2 (1)

$$\begin{pmatrix} x \\ y \\ z \end{pmatrix}$$

V. Ex. 2 (2) No decision

VI. (1)

(i)

$$\begin{pmatrix} z \\ y \\ x \end{pmatrix}$$

(ii)

$$\begin{pmatrix} z \\ x \\ y \end{pmatrix}$$

VI. (2) Preference structure (ii) is in favor of z.

VI. (3) In fact, the society prefers z above all, but there is a change in its preferences with regard to x and y, even though there is no change in its preferences with regard to them in the preference structure.

VI. (4) We do not recommend this rule, because the bias in favor of z will effect changes in subjects that are not relevant.

VII. (1)

(i)

$$\begin{pmatrix} y \\ z \\ x \end{pmatrix}$$

(ii)

$$\begin{pmatrix} y \\ z \\ x \end{pmatrix}$$

VII. (2) Everyone in structure (ii) prefers x to y.

VII. (3) There is no difference in society's preference.

VII. (4) We do not recommend this rule, because even though everyone prefers x to y, society prefers y to x.

2.6

1. The unanimous decision axiom

2. (1) No decision

2. (2) Axiom 1, according to which a social preference exists for every preference profile.

3. (1)

$$\begin{pmatrix} x \\ y \\ z \sim t \end{pmatrix}$$

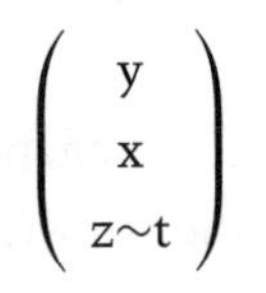

3. (2) The unanimous decision axiom

4. (1)

$$\begin{pmatrix} z \\ y \\ x \end{pmatrix}$$

$$\begin{pmatrix} y \\ x \\ z \end{pmatrix}$$

4. (2) The second profile is biased in favor of y.
4. (3) The independence of irrelevant alternatives axiom

5. The non-dictatorship axiom

2.8

1. (1) No decision is possible.
1. (2) Still no decision.
1. (3) There is not enough information about couple y, z and/or couple t, z.

2. (1) The social decision cannot be predicted.
2. (2) There still is not enough information.
2. (3)

$$\begin{pmatrix} x \\ y \\ z \end{pmatrix}$$

3. (1) No decision

3. (2)

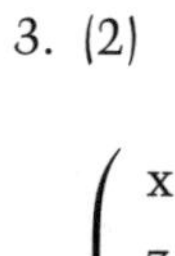

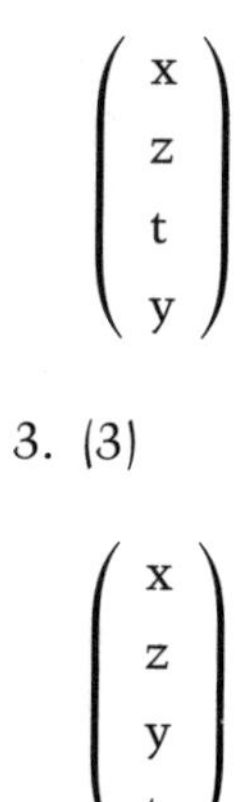

$$\begin{pmatrix} x \\ z \\ t \\ y \end{pmatrix}$$

3. (3)

$$\begin{pmatrix} x \\ z \\ y \\ t \end{pmatrix}$$

2.11

1. (1) No decision
1. (2)

$$\begin{pmatrix} x \\ t \\ z \\ y \end{pmatrix}$$

2. (1) No decision
2. (2)

$$\begin{pmatrix} x \sim z \\ y \end{pmatrix}$$

2. (3)

$$\begin{pmatrix} y \\ x \\ z \end{pmatrix}$$

2. (4) No decision

3. (1) i. No decision. Axiom 1 is not satisfied.

3. (1) ii.

$$\begin{pmatrix} x \\ y \\ z \sim t \end{pmatrix}$$

3. (1) iii.

$$\begin{pmatrix} y \\ x \\ z \sim t \end{pmatrix}$$

3. (2) Axiom 2 is not satisfied.

4.

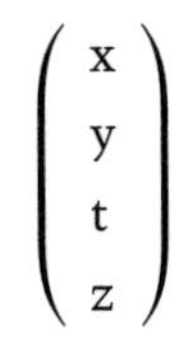

5. (1) No decision
5. (2) There is not enough information regarding couple x, y.

6. (1) No
6. (2) No
6. (3)

$$\begin{pmatrix} x \\ y \\ z \\ t \end{pmatrix}$$

A.3 CHAPTER 3

3.4

1. $v(1) = 2000$ $\quad v(2) = v(3) = 0$ $\quad v(\emptyset) = 0$
   $v(2,3) = 0$ $\quad v(1,2) = 2800$ $\quad v(1,3) = 3000$
   $v(1,2,3) = 3000$

2. $v(1) = v(2) = v(3) = 0$

   $v(1,2) = 0 \qquad v(1,3) = v(2,3) = 50$

   $v(1,2,3) = 50$

3. $v(1) = v(2) = 100 \qquad v(3) = v(4) = 0$

   $v(1,2) = v(3,4) = 0$

   $v(1,3) = v(1,4) = v(2,3) = v(2,4) = 150$

   $v(1,2,4) = v(1,2,3) = 250$

   $v(2,3,4) = v(1,3,4) = 150$

   $v(1,2,3,4) = 300$

4. $v(1) = v(2) = v(3) = v(4) = v(5) = 0$

   $v(1,2) = v(1,3) = v(2,3) = v(4,5) = 0$

   $v(1,4) = v(1,5) = v(2,4) = v(2,5) = v(3,4) = v(3,5) = 100$

   $v(1,2,3) = 0$

   $v(1,2,4) = v(1,2,5) = v(1,3,4) = v(1,3,5) = v(2,3,4) =$

   $= v(2,3,5) = v(3,4,5) = v(1,4,5) = v(1,2,5) = 100$

   $v(1,2,3,4) = v(1,2,3,5) = 100$

   $v(1,2,4,5) = v(1,3,4,5) = v(2,3,4,5) = 200$

   $v(1,2,3,4,5) = 200$

3.8

1. (1)

   $v(1) = v(2) = v(3) = 0$

   $v(2,3) = 0 \qquad v(1,2) = v(1,3) = v(1,2,3) = 1$

1. (2)

   $v(1) = v(2) = v(3) = v(4) = 0$

   $v(2,3) = v(2,4) = v(3,4) = 0$

   $v(1,2) = v(1,3) = v(1,4) = v(1,2,3) = v(1,2,4) =$

   $= v(1,3,4) = v(2,3,4) = v(1,2,3,4) = 1$

1. (3)

   $v(1) = v(2) = v(3) = 0$

   $v(1,2) = v(1,3) = v(2,3) = v(1,2,3) = 1$

1. (4)

$v(1) = v(2) = v(3) = 0$

$v(1,2) = v(1,3) = v(2,3) = v(1,2,3) = 1$

1. (5)

$v(1) = v(2) = v(3) = v(4) = 0$

$v(1,3) = v(2,3) = v(3,4) = 0$

$v(1,2) = v(1,4) = v(2,4) = 1$

$v(1,2,3) = v(1,2,4) = v(2,3,4) = v(1,3,4) = v(1,2,3,4) = 1$

1. (6)

$v(1) = 1 \qquad v(2) = v(3) = v(4) = 0$

$v(2,3,4) = 0 \qquad v(2,3) = v(2,4) = v(3,4) = 0$

$v(1,2) = v(1,3) = v(1,4) = 1$

$v(1,2,3) = v(1,2,4) = v(1,3,4) = v(1,2,3,4) = 1$

1. (7)

$v(1) = v(2) = v(3) = 0$

$v(1,2) = v(1,3) = v(2,3) = v(1,2,3) = 1$

1. (8)

$v(1) = v(2) = v(3) = v(4) = 0$

$v(1,4) = v(2,4) = v(3,4) = 0$

$v(1,2) = v(1,3) = v(2,3) = 1$

$v(1,2,3) = v(1,2,4) = v(2,3,4) = v(1,3,4) = v(1,2,3,4) = 1$

2. (1) $[5; 2, 2, 1, 1, 1]$

$v(1) = v(2) = v(3) = v(4) = v(5) = 0$

$v(1,2) = v(1,3) = v(1,4) = v(1,5) = v(2,3) =$

$= v(2,4) = v(2,5) = v(3,4) = v(3,5) = v(4,5) = 0$

$v(1,2,3) = v(1,2,4) = v(1,2,5) = 1$

$v(2,3,4) = v(2,3,5) = v(2,4,5) = v(3,4,5) = 0$

$v(1,3,4) = v(1,3,5) = v(1,4,5) = 0$

$v(1,2,3,4) = v(1,2,3,5) = v(1,3,4,5) =$

$= v(1,2,4,5) = v(2,3,4,5) = v(1,2,3,4,5) = 1$

2. (2) $[12; 5, 5, 3, 4]$

$v(1) = v(2) = v(3) = v(4) = 0$

$v(1,2) = v(1,3) = v(1,4) = v(2,3) = v(2,4) = v(3,4) = 0$

$$v(1,2,3) = v(1,2,4) = v(1,3,4) = v(2,3,4) = 1$$
$$v(1,2,3,4) = 1$$

2. (3) $[80; 50, 40, 30]$
$$v(1) = v(2) = v(3) = 0$$
$$v(2,3) = 0 \quad v(1,2) = v(1,3) = v(1,2,3) = 1$$

2. (4) $[80; 35, 35, 35, 15]$
$$v(1) = v(2) = v(3) = v(4) = 0$$
$$v(1,2) = v(1,3) = v(1,4) = v(2,3) = v(2,4) = v(3,4) = 0$$
$$v(1,2,3) = v(1,2,4) = v(1,3,4) = v(2,3,4) = 1$$
$$v(1,2,3,4) = 1$$

3. $v(1) = v(2) = v(3) = v(4) = 0$
$$v(1,2) = v(1,3) = v(1,4) = v(2,3) = v(2,4) = v(3,4) = 0$$
$$v(1,2,3) = v(1,2,4) = v(1,3,4) = 1 \qquad v(2,3,4) = 0$$
$$v(1,2,3,4) = 1$$

4. (1) Yes
4. (2) No
4. (3) Yes
4. (4) No
4. (5) Yes

3.10

1. (1) Players 1 and 2 are symmetric.
1. (2) All n players are symmetric.
1. (3) Players 1 and 2 are symmetric.
1. (4) All players are symmetric.
1. (5) Players 3 and 4 are symmetric.
1. (6) There are no symmetric players.

2. (1) All players are symmetric.
2. (2) Players 1 and 2 are symmetric.
2. (3) Players 2, 3, and 4 are symmetric.
2. (4) All players are symmetric.
2. (5) Players 1, 2, and 3 are symmetric.

3.12

1. (1) Players 1 is a null player.
1. (2) In this case, player 1 is not a null player.

2. (1) Players 1 and 2 are null players; players 3 and 4 are symmetric.
2. (2) In this case, players 1 and 2 are not null players.

3. (1) Player 4 is a null player.
3. (2) Player 4 is a null player.
3. (3) Players 4 and 5 are null players.

3.14

1.

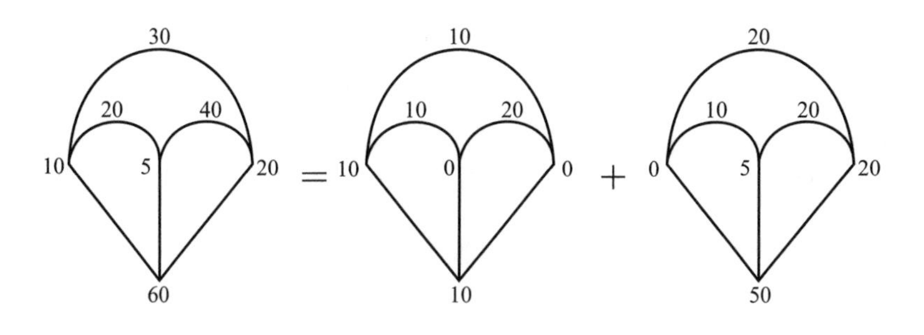

2. $(v+w)(1) = 5$ $\quad (v+w)(2) = 0$ $\quad (v+w)(3) = 5$
$(v+w)(1,2) = 20$ $\quad (v+w)(1,3) = 35$ $\quad (v+w)(2,3) = 30$
$(v+w)(1,2,3) = 70$

3. (1)

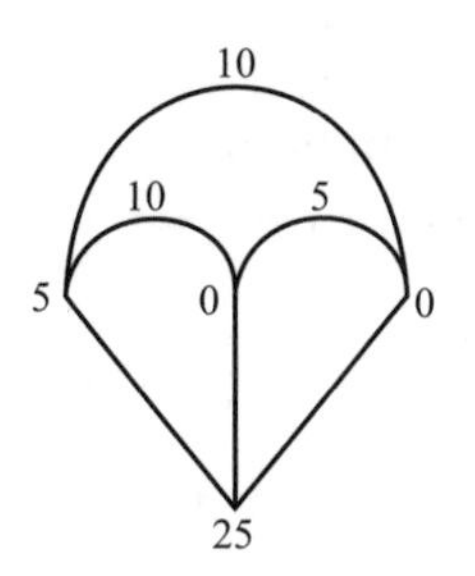

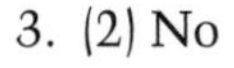

3. (2) No

4. (1)

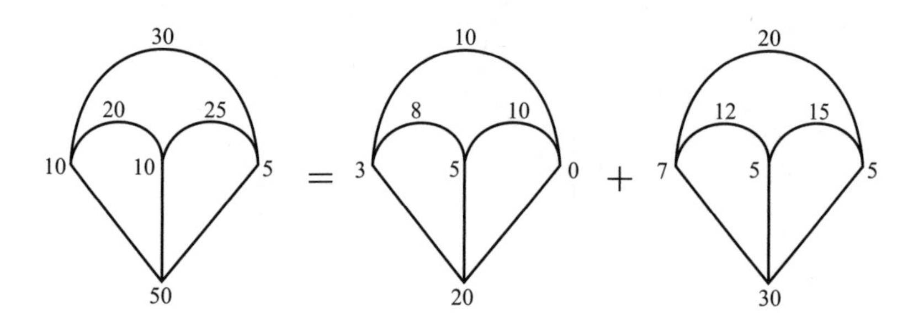

4. (2) No

5. (1)

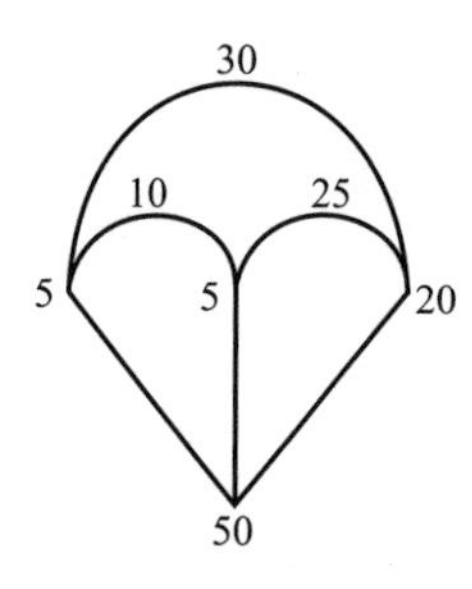

5. (2) No

6. (1)

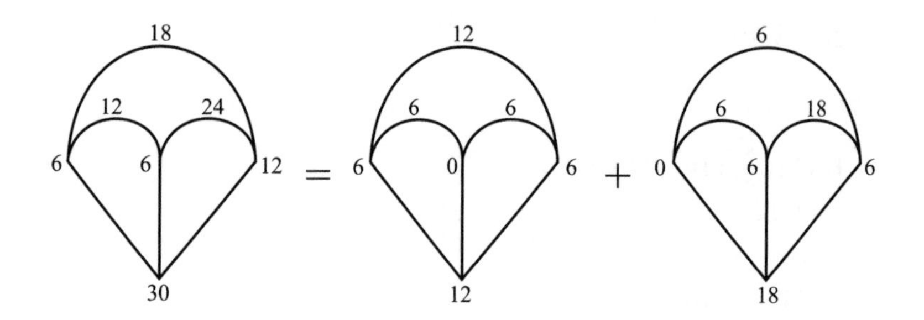

6. (2) Yes; players 2 and 3 are symmetric and 1 is a null player.

7.

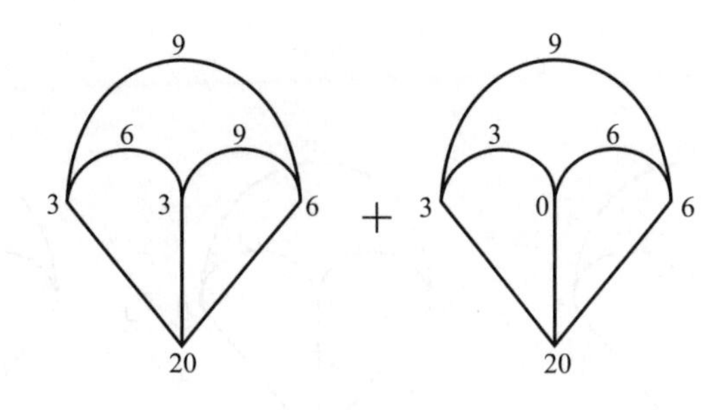

3.16

1. $\left(\frac{1}{2}, \frac{1}{2}\right)$

2. $\left(\frac{1}{3}, \frac{1}{3}, \frac{1}{3}\right)$

3. (1) $(1, 0, 0, 0)$
3. (2) $\left(\frac{1}{3}, \frac{1}{3}, \frac{1}{3}\right)$
3. (3) $\left(\frac{1}{3}, \frac{1}{3}, \frac{1}{3}, 0\right)$

4. $\left(116\frac{2}{3}, 116\frac{2}{3}, 66\frac{2}{3}\right)$

5. $(11, 14, 23)$

3.18

1. $\left(16\frac{2}{3}, 36\frac{2}{3}, 6\frac{2}{3}\right)$

2. $(27, 36, 9)$

3. $(74, 30, 62, 74)$

4. $(26, 14, 20)$

3.20

1. (1) $\left(116\frac{2}{3}, 116\frac{2}{3}, 66\frac{2}{3}\right)$
1. (2) $(3, 12, 9)$
1. (3) $\left(\frac{2}{3}, \frac{1}{6}, \frac{1}{6}\right)$
1. (4) $(11, 14, 23)$

2. $\left(\frac{c}{2}, \frac{c}{2}\right)$

3. (i) $\left(11\frac{5}{6}, 7\frac{1}{3}, \frac{5}{6}\right)$
3. (ii) $(5, 5, 0)$
3. (iii) $(6, 12, 12)$
3. (iv) $\left(6\frac{2}{3}, 6\frac{2}{3}, 6\frac{2}{3}, 0\right)$
3. (v) $(8, 8, 10, 10)$

3.22

1. $\left(\frac{1}{2}, \frac{1}{6}, \frac{1}{6}, \frac{1}{6}\right)$
2. $\left(\frac{1}{6}, \frac{1}{6}, \frac{1}{2}, \frac{1}{6}\right)$
3. $\left(\frac{1}{6}, \frac{1}{6}, \frac{1}{3}, \frac{1}{3}\right)$
4. $\left(\frac{2}{5}, \frac{3}{20}, \frac{3}{20}, \frac{3}{20}, \frac{3}{20}\right)$
5. $\left(\frac{3}{10}, \frac{3}{10}, \frac{2}{15}, \frac{2}{15}, \frac{2}{15}\right)$
6. $\left(\frac{13}{60}, \frac{13}{60}, \frac{17}{120}, \frac{17}{120}, \frac{17}{120}, \frac{17}{120}\right)$
7. $\left(\frac{3}{8}, \frac{5}{56}, \frac{5}{56}, \frac{5}{56}, \frac{5}{56}, \frac{5}{56}, \frac{5}{56}, \frac{5}{56}\right)$

3.24

1. $\left(\frac{9}{16}, \frac{7}{240}, \ldots, \frac{7}{240}\right)$
2. $\left(\frac{15}{26}, \frac{11}{650}, \ldots, \frac{11}{650}\right)$
3. $\left(\frac{45}{76}, \frac{31}{5700}, \ldots, \frac{31}{5700}\right)$
4. $\left(\frac{3}{11}, \frac{3}{11}, \frac{1}{22}, \ldots, \frac{1}{22}\right)$
5. $\left(\frac{11}{42}, \frac{11}{42}, \frac{1}{42}, \ldots, \frac{1}{42}\right)$
6. $\left(\frac{21}{82}, \frac{21}{82}, \frac{1}{82}, \ldots, \frac{1}{82}\right)$

3.26

1. $\frac{10! \cdot 8!}{2! \cdot 15!} = \frac{1}{143} \sim 0.7\%$

2. $\frac{6!\cdot 6!}{1!\cdot 11!} = \frac{1}{77} \sim 1.29\%$

3. $\frac{6!\cdot 5!}{11!} = \frac{1}{462} \sim 0.2\%$

3.28

1. $(15, 15, 20)$

2. $(6, 28, 30)$

3. $(6, 11, 11, 16, 17, 18)$

4. $(1400, 1400, 1400, 1400, 1400, 1500, 1500, 1500, 2100, 2100, 2100, 2100, 2100, 2100)$

5. (1) $c(L_1) = 120{,}000$ $c(L_1, L_2) = 120{,}000$ $c(L_1, H) = 200{,}000$ $c(M_1, M_2, H) = 200{,}000$

5. (2) $(200, 200, 350, 350, 350, 550)$

3.29

1. (1)
$v(1) = v(2) = v(3) = v(4) = 0$
$v(1,2) = 1$ $v(1,3) = 1$ $v(1,4) = 0$ $v(2,3) = 1$ $v(2,4) = 0$
$v(3,4) = 0$ $v(1,2,3) = 1$ $v(1,2,4) = 1$ $v(2,3,4) = 1$
$v(1,3,4) = 1$ $v(1,2,3,4) = 1$

1. (2) Yes. $X_1 = X_2 = X_3$.

1. (3) Yes. Player 4.

1. (4) $\left(\frac{1}{3}, \frac{1}{3}, \frac{1}{3}, 0\right)$

2. (1)

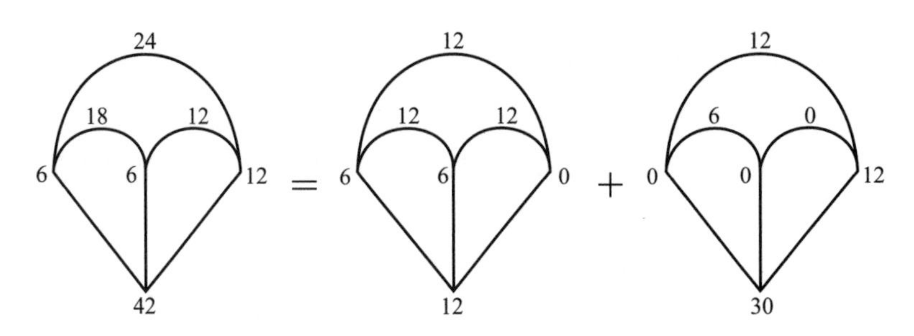

2. (2) No

3. $\left(\frac{3}{4}, \frac{1}{12}, \frac{1}{12}, \frac{1}{12}\right)$

4. $\left(58\frac{1}{3}, 58\frac{1}{3}, 33\frac{1}{3}\right)$

5. $(16, 19, 25)$

6. $\left(\frac{1}{3}, \frac{1}{6}, \frac{1}{4}, \frac{1}{4}\right)$

7. (2) $(8, 14, 20)$

8. $\left(\frac{11}{60}, \frac{11}{60}, \frac{17}{120}, \frac{17}{120}, \frac{17}{120}, \frac{17}{120}\right)$

A.4 CHAPTER 4

4.3

1. $\left(87\frac{1}{2}, 62\frac{1}{2}\right)$

2. $(130, 70)$

3. $(100, 100)$

4. $(210, 90)$

5. No; $(200, 300)$

6. $(125, 75)$

7. Yes

4.7

1. Yes

2. Yes

3. Yes

4. Yes. The widows with marriage contracts of 200 and 300 units, respectively.

5. The couple with marriage contracts of 200 and 300 units, respectively.

6. Yes

7. Yes

8. The creditors with claims of 150 and 300 units, respectively.

9. Yes

10. No

4.10

1. $\left(66\frac{2}{3}, 166\frac{2}{3}, 266\frac{2}{3}\right)$

2. $\left(33\frac{1}{3}, 83\frac{1}{3}, 183\frac{1}{3}\right)$

3. $(25, 50, 75, 80)$

4. $\left(40, 63\frac{1}{3}, 103\frac{1}{3}, 143\frac{1}{3}\right)$

5. $(25, 55, 105, 155, 205, 255)$

6. $(35, 50, 80, 110, 125)$

8. No

9. The division in the second and fifth columns in the table are inconsistent.

4.12

1. (1) $\left(25, 108\frac{1}{3}, 158\frac{1}{3}, 258\frac{1}{3}\right)$

1. (2) Yes

2. (1) $(25, 40, 50, 70, 100, 115)$
2. (2) Yes

3. (1) $(50, 75, 123\frac{3}{4}, 183\frac{3}{4}, 223\frac{3}{4}, 243\frac{3}{4})$
3. (2) Yes

4.14

1. $(50, 75, 75, 75)$

2. $(50, 100, 125, 125)$

3. (1) $(100, 150, 180, 180, 180)$

4. (1) $(40, 60, 80, 110, 110)$

4.17

1. $\left(66\frac{2}{3}, 116\frac{2}{3}, 216\frac{2}{3}\right)$

2. $(50, 100, 150, 200)$

3. $v(1) = v(2) = v(3) = 0 \quad v(1,2) = 100$
   $v(1,3) = 200 \quad v(2,3) = 400 \quad v(1,2,3) = 500$
   $\left(83\frac{1}{3}, 183\frac{1}{3}, 233\frac{1}{3}\right)$

4. $\left(116\frac{2}{3}, 150, 183\frac{1}{3}, 250\right)$
   $v(1) = v(2) = v(3) = v(4) = 0 \quad v(1,2) = 0$
   $v(1,3) = 50 \quad v(2,3) = 100 \quad v(1,4) = 150 \quad v(2,4) = 200$
   $v(3,4) = 250 \quad v(1,2,3) = 300 \quad v(1,2,4) = 400$
   $v(1,3,4) = 450$
   $v(2,3,4) = 500 \quad v(1,2,3,4) = 700$

5. (1) $\left(20\frac{5}{6}, 37\frac{1}{2}, 62\frac{1}{2}, 79\frac{1}{6}\right)$

6. (1) $\left(35, 51\frac{2}{3}, 91\frac{2}{3}, 121\frac{2}{3}\right)$

6. (2) $v(1) = v(2) = v(3) = v(4) = 0$
   $v(1,2) = v(1,3) = v(1,4) = v(2,3) = 0$
   $v(2,4) = 20 \quad v(3,4) = 100 \quad v(1,2,3) = 20$
   $v(1,2,4) = 100 \quad v(1,3,4) = 180 \quad v(2,3,4) = 220$
   $v(1,2,3,4) = 300$

4.19

1. $(225, 275), (250, 250), \left(230\frac{10}{13}, 269\frac{3}{13}\right), (225, 275)$

2. $(100, 200, 300, 400), \left(200, 266\frac{2}{3}, 266\frac{2}{3}, 266\frac{2}{3}\right),$
   $\left(142\frac{6}{7}, 214\frac{2}{7}, 285\frac{5}{7}, 357\frac{1}{7}\right), \left(141\frac{2}{3}, 208\frac{1}{3}, 270\frac{5}{6}, 379\frac{1}{6}\right)$

3. $\left(25, 50, 93\frac{3}{4}, 143\frac{3}{4}, 218\frac{3}{4}, 268\frac{3}{4}\right)$,
$\left(50, 100, 150, 166\frac{2}{3}, 166\frac{2}{3}, 166\frac{2}{3}\right)$,
$\left(38\frac{2}{21}, 76\frac{4}{21}, 114\frac{6}{21}, 152\frac{8}{21}, 190\frac{10}{21}, 228\frac{12}{21}\right)$

4. (1) $(100, 200, 200, 200)$, $(100, 200, 200, 200)$,
$\left(77\frac{7}{9}, 155\frac{5}{9}, 194\frac{4}{9}, 272\frac{2}{9}\right)$

5. $v(1) = 100$ $v(2) = 200$ $v(3) = 300$ $v(1, 2) = 400$
$v(1, 3) = 500$ $v(2, 3) = 600$ $v(1, 2, 3) = 800$
$\left(166\frac{2}{3}, 266\frac{2}{3}, 366\frac{2}{3}\right)$

Bibliography

Arrow, K. J. 1951. *Social choice and individual values.* New York: J. Wiley

Aumann, R. J. and Maschler, M. 1985. "Game-theoretic analysis of a bankruptcy problem from the Talmud," *Journal of Economic Theory* 36: 195–213

Gale, D. 2001. "The two-sided matching problem: origin, development and current issues," *International Game Theory Review* 3: 237–52

Gale, D. and Shapley, L. S. 1962. "College admissions and the stability of marriage," *American Mathematical Monthly* 69: 9–15

Harsanyi, J. C. 1959. "A bargaining model for the cooperative *n*-person game," in Tucker, A. W. and Luce, R. D. (eds.), *Contributions to the theory of games IV*, Annals of Mathematics Studies 40. Princeton: Princeton University Press, pp. 325–55

Kaminski, M. M. 2000. "Hydraulic rationing," *Mathematical Social Sciences* 40: 131–55

Maschler, M. 1982. "The worth of a cooperative enterprise to each member," in Diestler, M., Furst, E. and Schwodiauer, G. (eds.), *Games, economic dynamics and time series analysis.* New York: Springer, pp. 67–73

Milnor, J. and Shapley, L. S. 1961. "Values of large games II: oceanic games," The Rand Corporation, Memorandum RM-2649

O'Neill, B. 1982. "A problem of rights arbitration from the Talmud," *Mathematical Social Sciences* 2: 345–71

Roth, A. E. and Sotomayor, M. 1990. *Two-sided matching: a study in game-theoretic modeling and analysis.* Cambridge: Cambridge University Press

Schmeidler, D. 1969. "The nucleolus of a characteristic function game," *SIAM Journal of Applied Mathematics* 17: 1163–70

Shapley, L. S. 1953. "A value for *n*-person games," in Kuhn, H. and Tucker, A. W. (eds.), *Contributions to the theory of games II.* Princeton: Princeton University Press, pp. 307–17

Shapley, L. S. and Shubik, M. 1954. "A method for evaluation of the distribution of power in a committee system," *The American Political Science Review* 48: 787–92

Sobolev, A. I. 1975. "The characterization of optimality principles in cooperative games by functional equations," in Vorobiev, N. N. (ed.), *Matematicheskie Metody v Socialnix Naukax* 6. Academy of Sciences of the Lithuanian S. S. R. Vilnius, pp. 94–151

Index